**Notice**

This document is disseminated under the sponsorship of the Department of Transportation in the interest of information exchange. The United States Government assumes no liability for its contents or use thereof.

**Notice**

The United States Government does not endorse products or manufacturers. Trade or manufacturers' names appear herein solely because they are considered essential to the objective of this report.

| REPORT DOCUMENTATION PAGE | | Form Approved OMB No. 0704-0188 |
|---|---|---|
| Public reporting burden for this collection of information is estimated to average 1 hour per response, including the time for reviewing instructions, searching existing data sources, gathering and maintaining the data needed, and completing and reviewing the collection of information. Send comments regarding this burden estimate or any other aspect of this collection of information, including suggestions for reducing this burden, to Washington Headquarters Services, Directorate for Information Operations and Reports, 1215 Jefferson Davis Highway, Suite 1204, Arlington, VA 22202-4302, and to the Office of Management and Budget, Paperwork Reduction Project (0704-0188), Washington, DC 20503. | | |
| 1. AGENCY USE ONLY (Leave blank) | 2. REPORT DATE<br>January 2006 | 3. REPORT TYPE AND DATES COVERED<br>Interim report<br>October 2003 - September 2005 |
| 4. TITLE AND SUBTITLE<br>Detroit Deicing Decision Support Tool: Description, Operation, and Simulation Results | | 5. FUNDING NUMBERS |
| 6. AUTHOR(S)<br>Jonathan T. Lee | | |
| 7. PERFORMING ORGANIZATION NAME(S) AND ADDRESS(ES)<br>John A. Volpe National Transportation Systems Center<br>Research and Innovative Technology Administration<br>U.S. Department of Transportation<br>Cambridge, MA 02142 | | 8. PERFORMING ORGANIZATION REPORT NUMBER<br>DOT-VNTSC-NASA-05-01 |
| 9. SPONSORING/MONITORING AGENCY NAME(S) AND ADDRESS(ES)<br>Glean Research Center<br>National Aeronautics and Space Administration<br>Hampton, Virginia 23681-0001 | | 10. SPONSORING/MONITORING AGENCY REPORT NUMBER |
| 11. SUPPLEMENTARY NOTES | | |
| 12a. DISTRIBUTION/AVAILABILITY STATEMENT<br>This document is available to the public through the National Technical Information Service, Springfield, Virginia 22161. | | 12b. DISTRIBUTION CODE |
| 13. ABSTRACT (Maximum 200 words)<br>The John A. Volpe National Transportation Systems Center, sponsored by the National Aeronautics and Space Administration, developed a deicing decision support tool, for Detroit Metropolitan Wayne County Airport (DTW).[1] The deicing decision support tool (DST) estimates the amount of time an aircraft takes to go through deicing. This timing information enables the airport, the airlines, and air traffic controllers to work together to minimize delays and cancellation of flights by optimizing the airport and flight resources. This report discloses all the information relating to the DST.<br><br>[1] Patent pending. | | |
| 14. SUBJECT TERMS<br>Air traffic controllers, deicing, aircraft, decision support tool, weather-related delays; deicing pads | | 15. NUMBER OF PAGES<br>94 |
| | | 16. PRICE CODE |
| 17. SECURITY CLASSIFICATION OF REPORT<br>Unclassified | 18. SECURITY CLASSIFICATION OF THIS PAGE<br>Unclassified | 19. SECURITY CLASSIFICATION OF ABSTRACT<br>Unclassified | 20. LIMITATION OF ABSTRACT |

NSN 7540-01-280-5500

Standard Form 298 (Rev. 2-89)
Prescribed by ANSI Std. 239-18
298-102

# Acknowledgments

This work is supported by National Aeronautics and Space Administration Langley Research Center, under contract DTRS57-03-C-10014. The author would like to thank Northwest Airlines System Operations Control Center (Robert Muhs, Lorne Cass, Tim Reid, Jay Friedman, and Mark Gross) and Detroit Metropolitan Wayne County Airport ramp tower personnel (Laura Vaughn, Jeff Gordon, Vince Hopkins, and Daniel Vella) for their generous assistance in gathering the necessary information and for providing insightful comments. The author would also like to acknowledge the late Jack Perkins, of the Volpe National Transportation Systems Center (Volpe Center), for spearheading this particular project and George Skaliotis, also of the Volpe Center, for his suggestions during the course of the project. In addition, the author would like to thank C.H. Chen from George Mason University for his insightful comments on the model.

# METRIC/ENGLISH CONVERSION FACTORS

## ENGLISH TO METRIC

### LENGTH (APPROXIMATE)
1 inch (in) = 2.5 centimeters (cm)
1 foot (ft) = 30 centimeters (cm)
1 yard (yd) = 0.9 meter (m)
1 mile (mi) = 1.6 kilometers (km)

### AREA (APPROXIMATE)
1 square inch (sq in, in$^2$) = 6.5 square centimeters (cm$^2$)
1 square foot (sq ft, ft$^2$) = 0.09 square meter (m$^2$)
1 square yard (sq yd, yd$^2$) = 0.8 square meter (m$^2$)
1 square mile (sq mi, mi$^2$) = 2.6 square kilometers (km$^2$)
1 acre = 0.4 hectare (he) = 4,000 square meters (m$^2$)

### MASS - WEIGHT (APPROXIMATE)
1 ounce (oz) = 28 grams (gm)
1 pound (lb) = 0.45 kilogram (kg)
1 short ton = 2,000 pounds (lb) = 0.9 tonne (t)

### VOLUME (APPROXIMATE)
1 teaspoon (tsp) = 5 milliliters (ml)
1 tablespoon (tbsp) = 15 milliliters (ml)
1 fluid ounce (fl oz) = 30 milliliters (ml)
1 cup (c) = 0.24 liter (l)
1 pint (pt) = 0.47 liter (l)
1 quart (qt) = 0.96 liter (l)
1 gallon (gal) = 3.8 liters (l)
1 cubic foot (cu ft, ft$^3$) = 0.03 cubic meter (m$^3$)
1 cubic yard (cu yd, yd$^3$) = 0.76 cubic meter (m$^3$)

### TEMPERATURE (EXACT)
[(x-32)(5/9)] °F = y °C

## METRIC TO ENGLISH

### LENGTH (APPROXIMATE)
1 millimeter (mm) = 0.04 inch (in)
1 centimeter (cm) = 0.4 inch (in)
1 meter (m) = 3.3 feet (ft)
1 meter (m) = 1.1 yards (yd)
1 kilometer (km) = 0.6 mile (mi)

### AREA (APPROXIMATE)
1 square centimeter (cm$^2$) = 0.16 square inch (sq in, in$^2$)
1 square meter (m$^2$) = 1.2 square yards (sq yd, yd$^2$)
1 square kilometer (km$^2$) = 0.4 square mile (sq mi, mi$^2$)
10,000 square meters (m$^2$) = 1 hectare (ha) = 2.5 acres

### MASS - WEIGHT (APPROXIMATE)
1 gram (gm) = 0.036 ounce (oz)
1 kilogram (kg) = 2.2 pounds (lb)
1 tonne (t) = 1,000 kilograms (kg)
            = 1.1 short tons

### VOLUME (APPROXIMATE)
1 milliliter (ml) = 0.03 fluid ounce (fl oz)
1 liter (l) = 2.1 pints (pt)
1 liter (l) = 1.06 quarts (qt)
1 liter (l) = 0.26 gallon (gal)



1 cubic meter (m$^3$) = 36 cubic feet (cu ft, ft$^3$)
1 cubic meter (m$^3$) = 1.3 cubic yards (cu yd, yd$^3$)

### TEMPERATURE (EXACT)
[(9/5) y + 32] °C = x °F

## QUICK INCH - CENTIMETER LENGTH CONVERSION

## QUICK FAHRENHEIT - CELSIUS TEMPERATURE CONVERSION

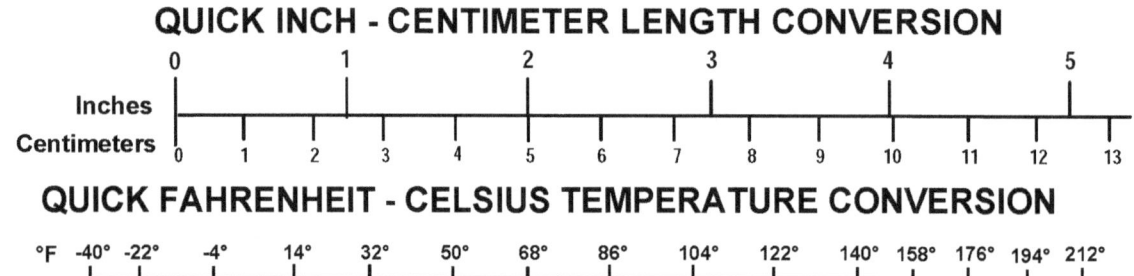

For more exact and or other conversion factors, see NIST Miscellaneous Publication 286, Units of Weights and Measures. Price $2.50 SD Catalog No. C13 10286

# Table of Contents

1. Introduction ..................................................................................................................1
   1.1 Motivation ............................................................................................................1
   1.2 Background ..........................................................................................................1
   1.3 Problem Statement ...............................................................................................2
2. Deicing Decision Support Tool ....................................................................................5
   2.1 Queuing Model ....................................................................................................5
   2.2 Parameters ............................................................................................................7
   2.3 MATLAB Code ....................................................................................................8
      2.3.1 Variable Names ..............................................................................................9
      2.3.2 Explanation of the Code ..............................................................................10
3. User Manual ...............................................................................................................13
   3.1 Overview ............................................................................................................13
   3.2 Installation .........................................................................................................13
      3.2.1 System Requirements ...................................................................................13
      3.2.2 DST Setup ....................................................................................................13
   3.3 Operations ..........................................................................................................14
      3.3.1 Deicing Pad Status .......................................................................................16
      3.3.2 Input Type of Aircraft in the Pad .................................................................17
      3.3.3 Input Type of Aircraft in the Queue .............................................................18
      3.3.4 Input Snow Type ..........................................................................................19
      3.3.5 Type of Outbound Aircraft ..........................................................................21
      3.3.6 Run Simulation ............................................................................................22
   3.4 Customization ....................................................................................................22
      3.4.1 Configure Deicing Pads ...............................................................................23
      3.4.2 Configure Pad Time .....................................................................................24
      3.4.3 Wide-Body Deicing .....................................................................................25
   3.5 Disclaimer ..........................................................................................................25
4. Results ........................................................................................................................27
   4.1 Snow Type ..........................................................................................................27
   4.2 Aircraft Type ......................................................................................................28
   4.3 Nonlinearity .......................................................................................................29

| | | |
|---|---|---|
| 5. | Future Directions | 33 |
| 7. | Contact Information | 35 |
| References | | 37 |
| Appendix A. | Volpe Project Team | 39 |
| Appendix B. | Detroit Deicing Decision Support Tool, Build 2, MATLAB Code | 41 |

# List of Figures

Figure 1. Diagram of DTW ................................................................................................3

Figure 2. The Deicing Problem .........................................................................................3

Figure 3. Deicing Queuing Model .....................................................................................6

Figure 4. DTW Deicing Pad 4R ........................................................................................6

Figure 5. DTW Deicing Pad 3L .........................................................................................7

Figure 6. Deicing Decision Support Tool, Build 2 ............................................................9

Figure 7. Deicing Tool Flow Chart ..................................................................................12

Figure 8. Detroit Deicing Decision Support Tool ...........................................................14

Figure 9. Text Display Window ......................................................................................15

Figure 10. Deicing Pad Status..........................................................................................16

Figure 11. Aircraft Type in the Deicing Pads..................................................................17

Figure 12. Aircraft Type in Queue...................................................................................18

Figure 13. Snow Type......................................................................................................19

Figure 14. Manual Pad Time Input..................................................................................20

Figure 15. Outbound A/C Type .......................................................................................21

Figure 16. Start Simulation..............................................................................................22

Figure 17. Pad Configuration Window............................................................................23

Figure 18. Warning Window ...........................................................................................24

Figure 19. Pad Time Configuration Window ..................................................................24

Figure 20. Error Window.................................................................................................25

Figure 21. Type A Snow..................................................................................................27

Figure 22. Type C Snow, Outbound A320 ......................................................................28

Figure 23. Outbound B757 ..............................................................................................29

Figure 24. Two B757s at Front of Queue ................................................................................ 30

Figure 25. Two A320s at Front of Queue ................................................................................ 30

# List of Tables

Table 1. Deicing Pad Configurations ........................................................................................ 7

Table 2. Mean Pad Time ........................................................................................................... 8

Table 3. Description of Snow Type .......................................................................................... 8

# List of Acronyms

| | |
|---|---|
| A/C | Aircraft |
| ATIDS | Airport Target Identification System |
| DROMS | Dynamic Runway Occupancy Measurement System |
| DST | Decision Support Tool |
| DTW | Detroit Metropolitan Wayne County Airport |
| GUI | Graphical User Interface |
| MC | Monte Carlo simulation |
| MEM | Memphis International Airport |
| NAS | National Airspace System |
| NASA | National Aeronautics and Space Administration |
| NWA | Northwest Airlines |

# Executive Summary

Smooth and efficient operation of the National Airspace System depends on timely execution of flight-related events. Weather can severely disrupt the carefully planned flight schedules at a hub airport and impact travelers throughout the country. In particular, a snowstorm may cause substantial perturbation in the departure of aircraft due to the need for deicing prior to take off. The additional time needed for an aircraft to be deiced, including time in queue, is highly nonlinear and difficult to predict. The Volpe National Transportation Systems Center, sponsored by the National Aeronautics and Space Administration, developed a deicing decision support tool for Detroit Metropolitan Wayne County Airport.[2] The Deicing Decision Support Tool (DST) estimates the amount of time an aircraft takes to go through deicing. This timing information enables the airport, the airlines, and air traffic controllers to work together to minimize delays and cancellation of flights by optimizing airport and flight resources. This report discloses all the information relating to the DST.

---

[2] Patent pending.

# 1. Introduction

## 1.1 Motivation

Delays associated with the departure phase of a flight account for the majority of delays in the National Airspace System (NAS) [4]. Moreover, the bulk of the delays are weather related. Snow or freezing rain is one possible cause of weather-related delays. For airports that experience long winter months; e.g., Detroit Metropolitan Wayne County Airport (DTW), deicing is an unavoidable part of operations. If the deicing resources are not managed well, there could be significant impact on the departure schedules. In addition, the flight schedules at hub airports are highly coupled to facilitate the passenger connections and crew exchanges at those airports. Any perturbation to the schedules could have unexpected impacts on the downstream connecting flights [1]. If it is snowing at a hub airport, deicing associated delays at the hub airport can be amplified by the hub-and-spoke system. This project addresses this issue by development of an automated tool, the Detroit Deicing Decision Support Tool, to help assure efficient use of deicing pads at DTW.[3] In particular, the John A. Volpe National Transportation Systems Center (Volpe Center) supported the National Aeronautics and Space Administration (NASA) in developing a deicing decision support tool for a hub airport.

This report summarizes the technical information on the deicing tool. In this section, the background and the problem statement are present. The technical details of the tool can be found in Section 2. The operations of the tool are depicted in Section 3. Section 4 contains the simulation results from the tool. Section 5 ends with a discussion of possible future directions for the tool. Appendix A lists the members on the Volpe Team. Appendix B contains the MATLAB code associated with the deicing decision support tool.

## 1.2 Background

The Volpe Center supported NASA in studying the benefits of the Airport Target Identification System (ATIDS) at DTW during 2003. ATIDS is a multilateration system, which provides real-time surface surveillance data at DTW. One of the benefits of ATIDS is the resolution of systematic surface flow problems [3]. An example is the imbalance of the loading of deicing pads during the deicing operations. One of the pads had a large queue, while the other one was almost empty.

In addition, the Volpe Center assisted in the development of the Dynamic Runway Occupancy Measurement System (DROMS). DROMS collects surface surveillance, weather, and airlines flight data and computes the runaway occupancy times for all arriving and departing aircraft

---

[3] Patent pending.

(A/C) and stores them in a database. The Volpe Center analyzed the relationship of different parameters such as aircraft type, weather, airport configuration, etc., between runaway occupancy times of arrival and departure for Memphis International Airport (MEM) and at DTW.

In November 2003, the Volpe Center proposed a number of possible applications of DROMS to Northwest Airlines (NWA), the major operator at DTW. One of the proposals was to develop a tool to estimate the total system, the amount of time for an aircraft to go through the deicing process, including the time in queue and the actual deicing time. NWA indicated that a deicing tool estimating the total system time would be the most beneficial application to the operators at DTW.

## 1.3 Problem Statement

There are two major deicing pads at DTW, next to runways 4R and 3L, as indicated in Figure 1. The problem is to estimate the total system time for an outbound aircraft that is to join either one of the two deicing queues. Figure 2 is a graphical representation of the problem statement. By providing the estimates for both pads, the tool would provide the user with information on the more efficient pad assignment for the particular outbound aircraft. This is similar to a shopper trying to determine which supermarket checkout line is faster. Besides assisting in the pad assignment, there are additional potential applications of this tool, which are discussed at the end of the report.

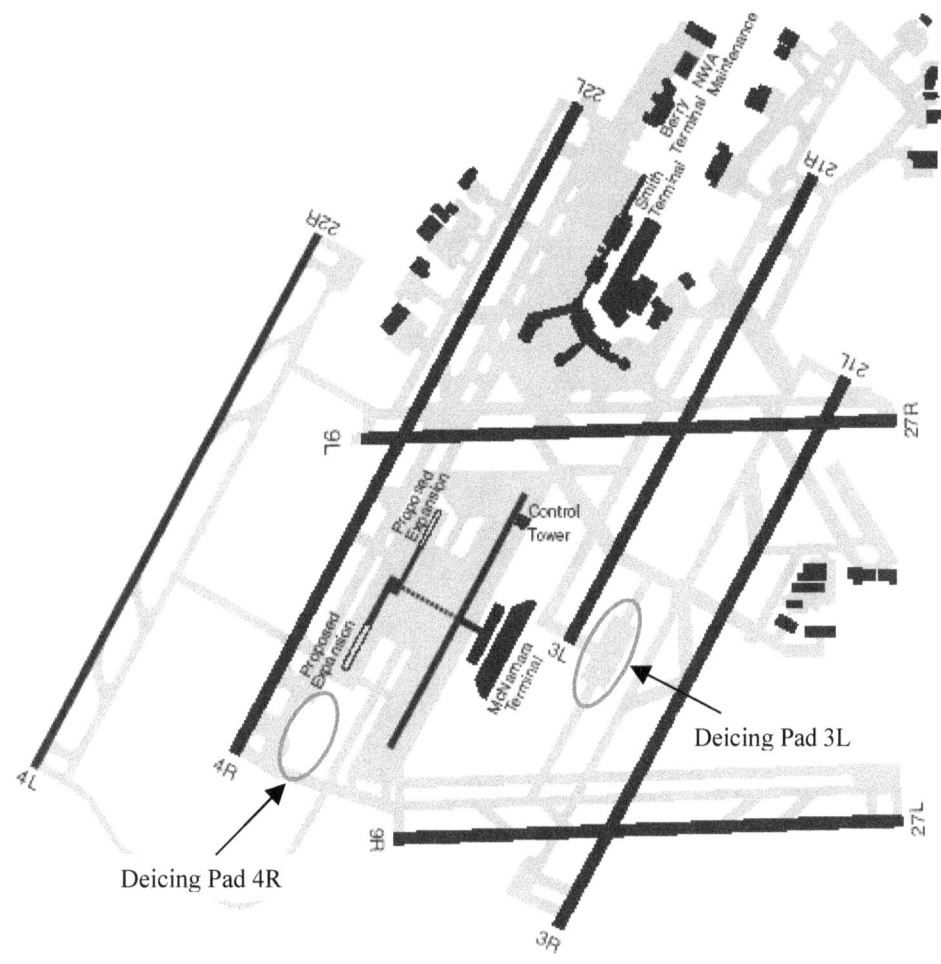

**Figure 1.** Diagram of DTW

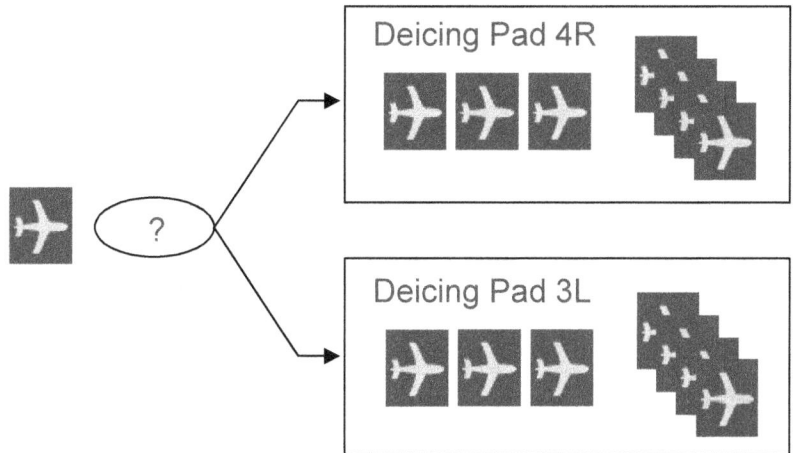

**Figure 2.** The Deicing Problem

# 2. Deicing Decision Support Tool

## 2.1 Queuing Model

At the heart of the deicing decision support tool is a queuing model, which mimics the process at a deicing pad. After consulting with NWA deicing experts, an initial queuing model was constructed, as shown in Figure 3. Monte Carlo simulations of the model estimate the total system time.

The queuing model captures all critical aspects of the deicing process, which are:
- Queue
- Control
- Server

As an aircraft departs the gate area, it first enters the *queue* of the assigned deicing pad. Each aircraft can be served at one of several deicing positions within the deicing pad. Once the aircraft is at the beginning of the queue, it awaits a deicing position that could accommodate the physical size of the aircraft. Conceptually, there is a *control* unit, which directs the leading aircraft in the queue to the appropriate open deicing position, between the queue and the pad. Subsequently, the aircraft goes to the *server*, which includes two processes: *taxiing* and *pad*. If a deicing position is open and can accommodate the leading aircraft in the queue, then the aircraft would *taxi* into the deicing position in the pad and be sprayed with deicing fluids. The time to taxi into position is called *taxi time*. The remaining time in the position referred to as *pad time* since the aircraft would physically be in the deicing pad waiting to be sprayed and being sprayed. If the open position cannot accommodate the leading aircraft in the queue, then the aircraft has to wait until a position that can accommodate it opens up. In turn, it would hold up every aircraft behind it in the queue.

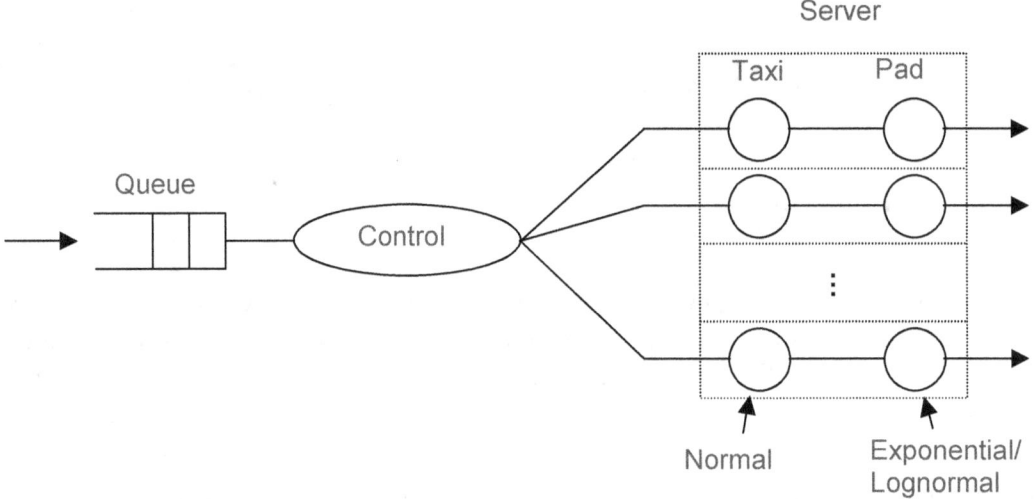

**Figure 3.  Deicing Queuing Model**

This queuing model can be described as *D/M, N-lognormal/m/n*. *D* denotes the deterministic interarrival time for the queue. In fact, the interarrival time is zero since all the aircraft are assumed to be in queue. *M* represents Markovian service time. That is only true for the aircraft that is already being deiced when the simulation commenced. Otherwise, the server has two serial stages with the following service times for the aircraft in the queue at the beginning of the simulation: normal/Gaussian (*N*) for taxi and *lognormal* for the pad time. (These distributions can be changed to different ones if supported by empirical data. Changes can be adapted to different airport operational procedures.) The letter *m* identifies the number of servers in the model. In other words, *m* is the number of deicing positions in a particular deicing pad. Currently, *m* is set to 7 but could be expanded in the future. The letter *n* represents the capacity of the queue, i.e., the maximum number of aircraft allowed in the queues. In the current version, *n* is set to 12.

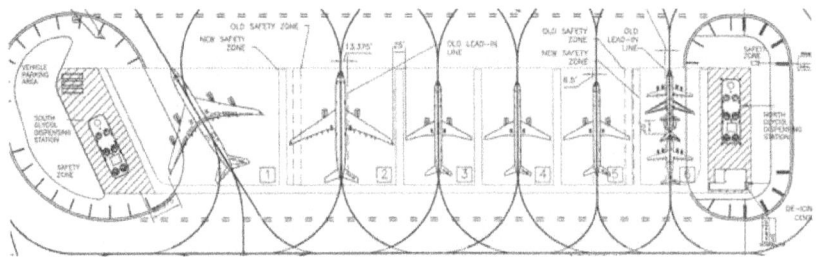

**Figure 4.  DTW Deicing Pad 4R**

The deicing pad configurations for Deicing Pad 4R and 3L can be found in Figure 4 and Figure 5, respectively. The physical constraints of each of the deicing positions within each of the pads can be found in Table 1.

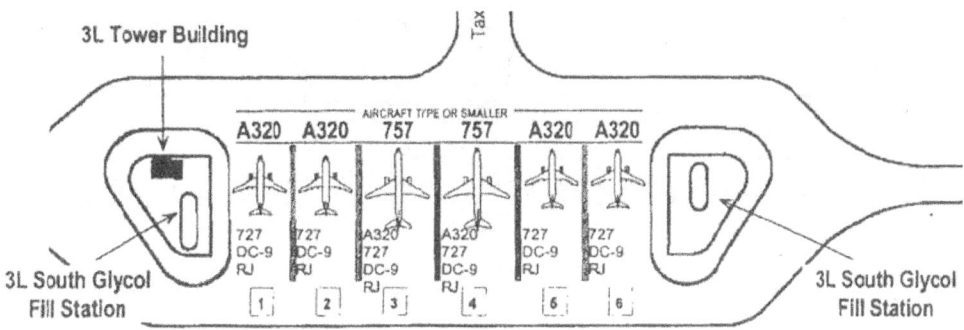

**Figure 5.** DTW Deicing Pad 3L

**Table 1. Deicing Pad Configurations**

|            | 4R   | 3L   |
|------------|------|------|
| Position 1 | B747 | A320 |
| Position 2 | A330 | A320 |
| Position 3 | B757 | B757 |
| Position 4 | B757 | B757 |
| Position 5 | B757 | A320 |
| Position 6 | DC9  | A320 |

## 2.2 Parameters

The queuing model is simulated as a timed state automaton. For more details on the timed state automata, please refer to Section 3.2 in [2]. The deicing queuing model is driven by the taxi and pad times. The taxi time is assumed to have a normal distribution with a mean of one minute and standard deviation of ten seconds for all aircraft type and in all snow types. The pad time is a stochastic process modeled either as an exponential random variable or as a lognormal random variable. For an aircraft that is already inside a deicing position being deiced at the beginning of the simulation, its pad time is modeled as an exponential random variable. This is to take advantage of the memory-less property of the exponential distributions. For an aircraft that is in the queue during the beginning of the simulation, the time it enters the deicing position would be known precisely during simulation. Thus, the pad time is modeled as a lognormal random variable. The lognormal distribution is chosen to match the historical pad times that have been observed. In fact, the lognormal distribution is supported in the positive domain, which mimics real-world observation, i.e., pad time is never less than zero minutes. The mean time used for both the exponential and the lognormal distributions can be found in Table 2.

**Table 2. Mean Pad Time**

|            | A  | B  | C  | D  | E  |
|------------|----|----|----|----|----|
| B747       | 18 | 23 | 30 | 40 | 60 |
| A330/DC10  | 13 | 18 | 23 | 34 | 45 |
| B757       | 8  | 11 | 16 | 20 | 45 |
| A319/A320  | 7  | 9  | 14 | 19 | 45 |
| DC9        | 3  | 5  | 12 | 15 | 45 |
| RJs        | 3  | 5  | 12 | 15 | 45 |

The mean pad time is a function of the aircraft type as well as the snow type. The larger the aircraft, the longer the pad time would be, in the same snow type. The snow type is also a variable that would influence the mean pad time. The snow type is a classification developed by Northwest Airlines (NWA) [5]. "A" is the least severe snow type while "E" is the most severe snow type. The more severe the snow type is, the longer the mean pad time would be. Descriptions of the snow types can be found in Table 3.

**Table 3. Description of Snow Type**

|   | |
|---|---|
| A | Frost, freezing fog or mist, light dry snow less than or equal to ¼"/HR, visibility ½ mile or greater |
| B | Rim ice, light wet snow less than or equal to ¼"/HR, visibility greater than or equal to ½ mile, moderate dry snow ¼ to ¾"/HR visibility greater than ¼ mile |
| C | Moderate wet snow greater than ¾"/HR visibility greater than ¼ mile, heavy dry snow greater than ¾"/HR, visibility less than or equal to ¼ mile |
| D | Heavy wet snow greater than ¾"/HR, visibility less than ¼ mile, light freezing drizzle, visibility less than ½ mile |
| E | Freezing rain: light, moderate, or heavy |

## 2.3 MATLAB Code

The deicing tool was developed using MATLAB. To enhance the usability of the tool, a graphical user interface (GUI) was added to the tool. Figure 6 is a screen shot of the GUI. Build 2 of the tool has been developed at the writing of this report. The complete MATLAB code for the Deicing DST, Build 2, can be found in Appendix B.

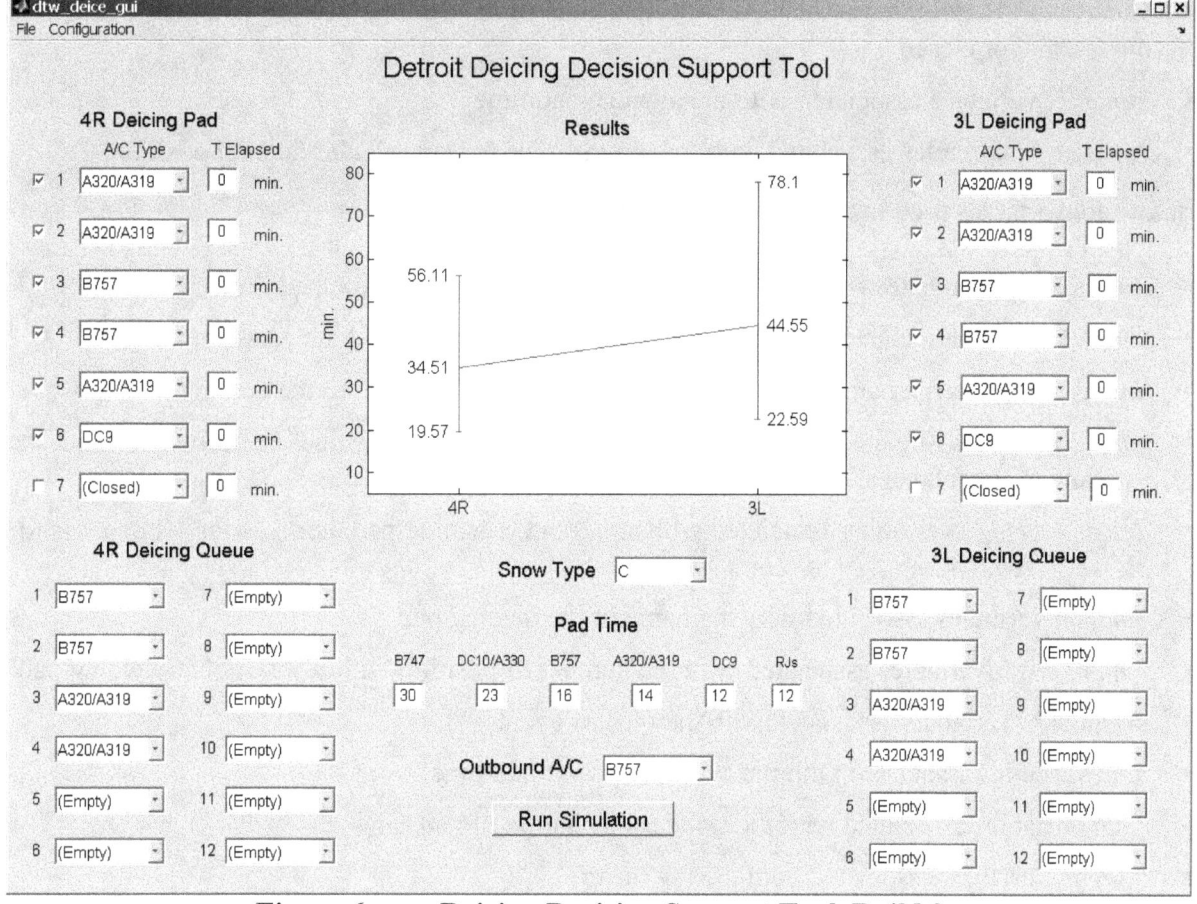

Figure 6.    Deicing Decision Support Tool, Build 2

## 2.3.1 Variable Names

There are many variables used in the program. Before getting into the details of the program, the following describes the convention used to name the variables. Variable names will be italicized. An asterisk (*) denotes a wildcard character.

Below are variables associated with GUI of the tool:

- *handleds.\**: variables associated with MATLAB GUI
- *\*type_\**: variables for the type of aircraft
- *\*_s_\**: variable for the deicing pad, i.e., the server
- *\*_q_\**: variables for the queue
- *\*_4_\**: variables associated with Deicing Pad 4R
- *\*_3_\**: variables associated with Deicing Pad 3L

- *checkbox_*: variables associated with the status of a particular deicing position, i.e., if it is operational or closed
- *time_*: variables associated with the average pad time
- *ac_list_*: variables associated with the aircraft type for a particular deicing position

Below are variables used in the queuing model:

- *dst*: variable associate with the type of distribution used to generate pad time
- *simul*: variable associated with the number of replication in the Monte Carlo simulation
- *dtw_deice_data*: file name associated with all the necessary input to the deicing tool
- *conf*: variable associated with the confidence interval that would be drawn around the estimate of the total system time
- *deice_pad(\*).\**: variables associated with a particular deicing pad, i.e., 1 is for 4R and 2 is for 3L
- *\*.name*: variables associated with the name of the deicing pad
- *\*.num_pos\**: variables associated with the number of position within a particular deicing pad
- *event(\*).\**: variables associated with the type of event
- *\*.m*: variables associated with the average time of pad time
- *\*.sd*: variables associated with the standard deviation of pad time
- *\*.logm*: variables associated with the log mean
- *\*.logsd*: variables associated with the log standard deviation
- *snow_type*: variable associated with the snow type under simulation
- *\*.size_res*: variables associated of the physical configuration of the deicing pad
- *\*.t*: variables associated with the clock in the queuing model

### 2.3.2 Explanation of the Code

A flow chart of the MATLAB code can be found in Figure 7. The program totals 2471 lines. The code from line 1 to line 1873 are all related to the GUI functions of the tool, correspond to steps 2 and 3 in flow chart. Line 1874 and onward are simulation related code, relate to steps 4 through 20. The first part is to initialize all the relevant variables in the simulation as well as the variables related to the outputs from the program. This is step 4. The initialization of the simulation takes place from line 1978 through line 2092. There are three nested loops. The first nested loop is from line 2095 to line 2401, match steps 6 through 19. This first loop would switch the simulation between the two deicing pads. The initialization of the simulation runs for each pad starts on line 2097 and ends on line 2119, correspond to step 7. The second nested loop accounts for the Monte Carlo simulations. The Monte Carlo simulation loop is from line 2125 through line 2390, match steps 9 through 17. A default value of 1,000 runs is set. An individual run of the queuing model resides in the innermost loop. Line 2127 is the beginning of the queuing model, which ends on line 2389, corresponds to step 11 through 17. After the runs are completed for one pad, statistical analysis is performed on the runs for that pad in lines 2391 through 2400, corresponding to step 18. When simulation runs for both pads are completed, the results of the simulation are displayed and described in the code from line 2407 onward, match step 20. Comments on the codes can be found throughout the program.

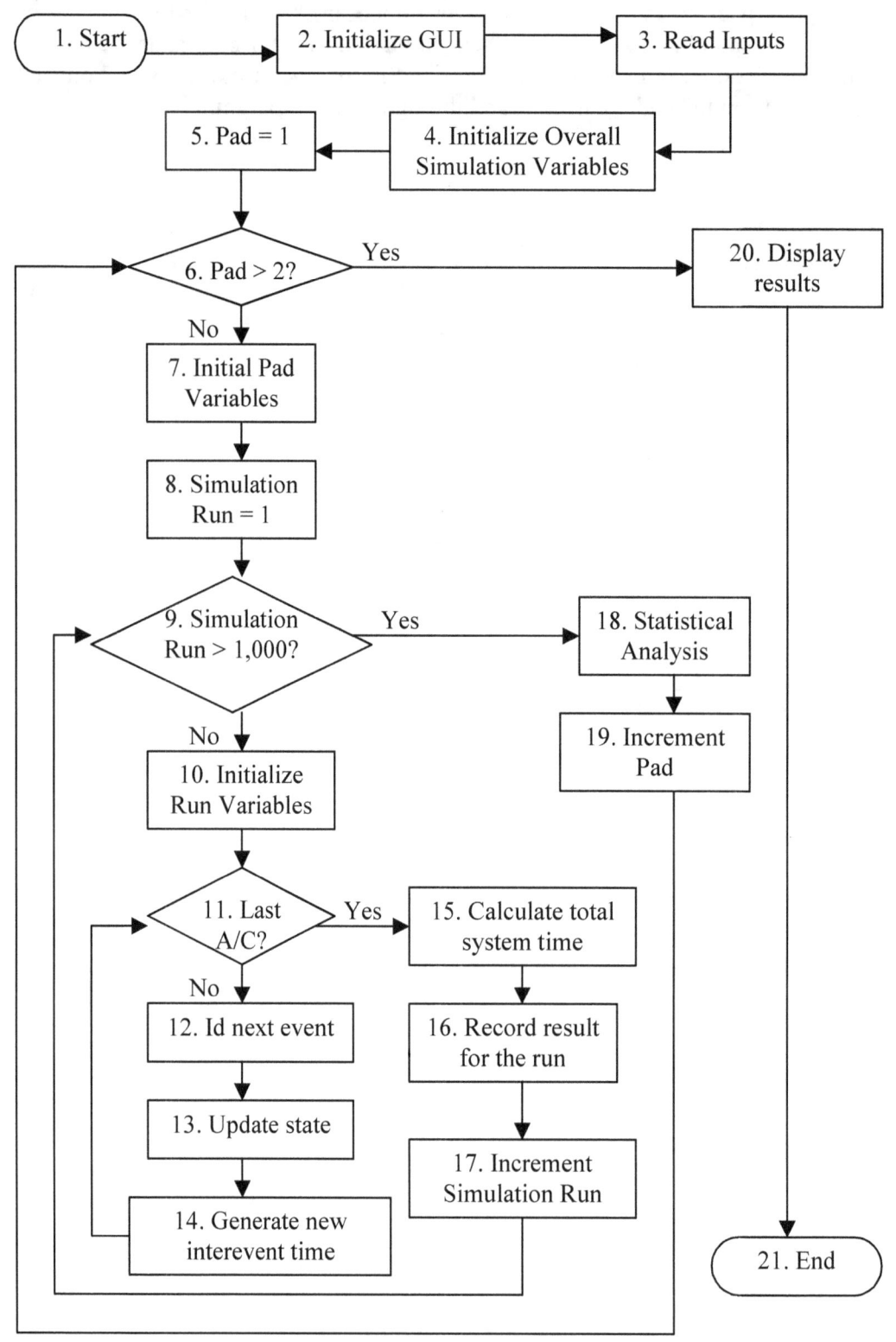

Figure 7. Deicing Tool Flow Chart

# 3. User Manual

## 3.1 Overview

This section of the report is intended as a user manual to the deicing decision support tool (DST), Build 2. The deicing DST is designed to provide information to assist in the load balancing of the deicing pads 4R and 3L at DTW. In particular, the tool estimates the amount of time an outbound aircraft needs to go through the deicing process, which includes waiting and being sprayed. The tool considers the current state of the deicing pads when predicting times for any particular aircraft.

The manual covers the installation process, operations, and customization of the tool. The Detroit Deicing Decision Support Tool was developed by the Volpe National Transportation Systems Center (Volpe Center), U.S. Department of Transportation. To obtain additional information, please refer to Section 6.

## 3.2 Installation

To install the tool, you will need to obtain the CD-ROM with all the necessary files, as explained in Section 6. Below are the details of the installation process.

### 3.2.1 System Requirements

To install the tool, your system will need to meet the following requirements:

1. Window-based computer system.
2. Administrative privileges on the computer system, i.e., you are an administrator on the computer system.
3. CD-ROM drive.
4. Mouse or compatible pointing device.
5. Permission may be needed from your network or information technology group to install this tool on the intended machine.

### 3.2.2 DST Setup

1. Close all programs.
2. Insert CD into the drive.
3. Copy MCRinstaller.exe from the CD into a temporary folder on the PC.
4. Run MCRinstaller.exe from the PC and follow the instructions from the prompt.
5. Create directory *Deicing* on the PC.

6. Upload files *dtw_deice_data.mat*, *dtw_deice_gui.ctf*, and *dtw_deice_gui.exe* from the CD into the directory *Deicing*.
7. Create a shortcut of the file *dtw_deice_gui.exe* on the desktop if desired.

If you have any difficulty installing the software, please contact your network or information technology group first. For additional assistance, refer to Section 6.

## 3.3 Operations

This section discusses the operations of the deicing DST. First, locate and click on *dts_deice_gui.exe* or the associated shortcut. Once the tool is running, you should see a display similar to that in Figure 8, along with a text display window similar to Figure 9. The text display window shows a text version of the simulation result. If desired, the text display window can be minimized.

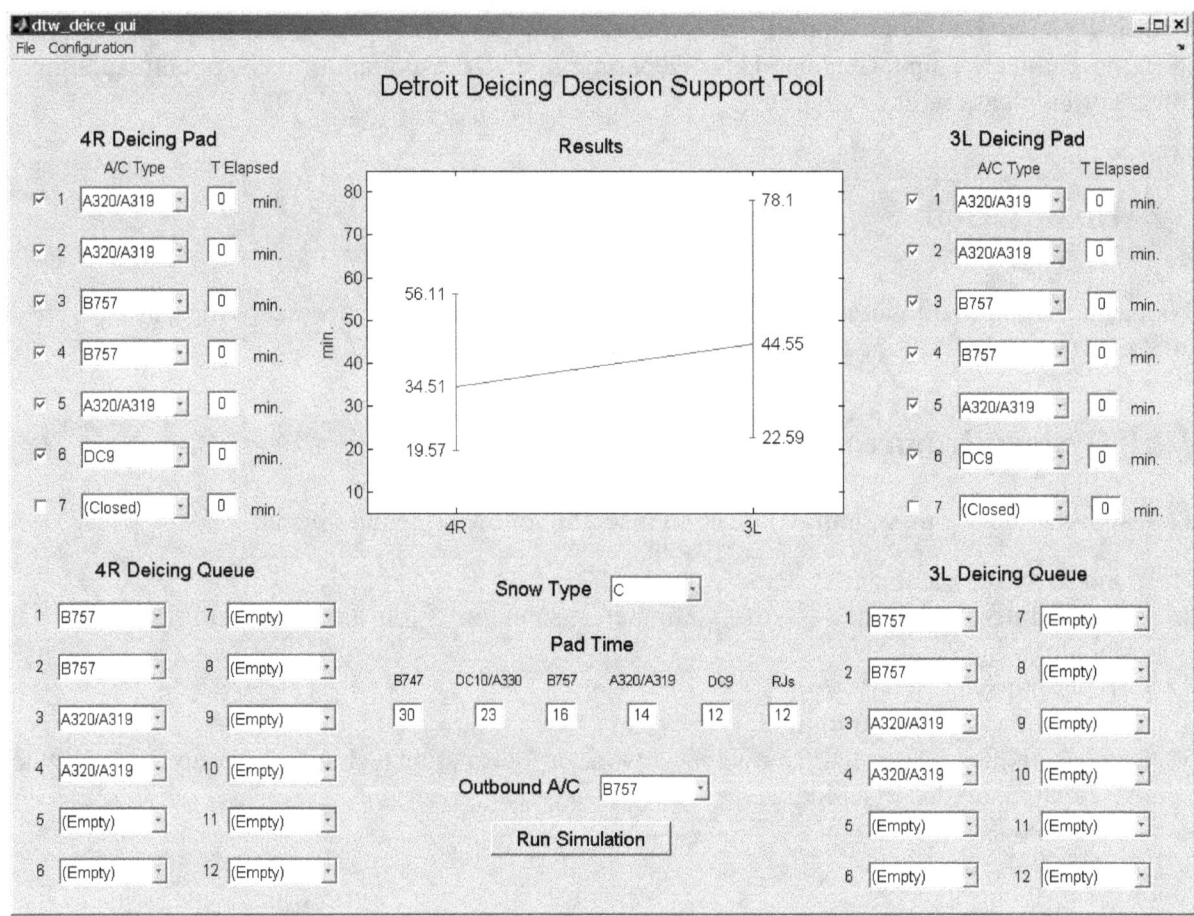

**Figure 8.** **Detroit Deicing Decision Support Tool**

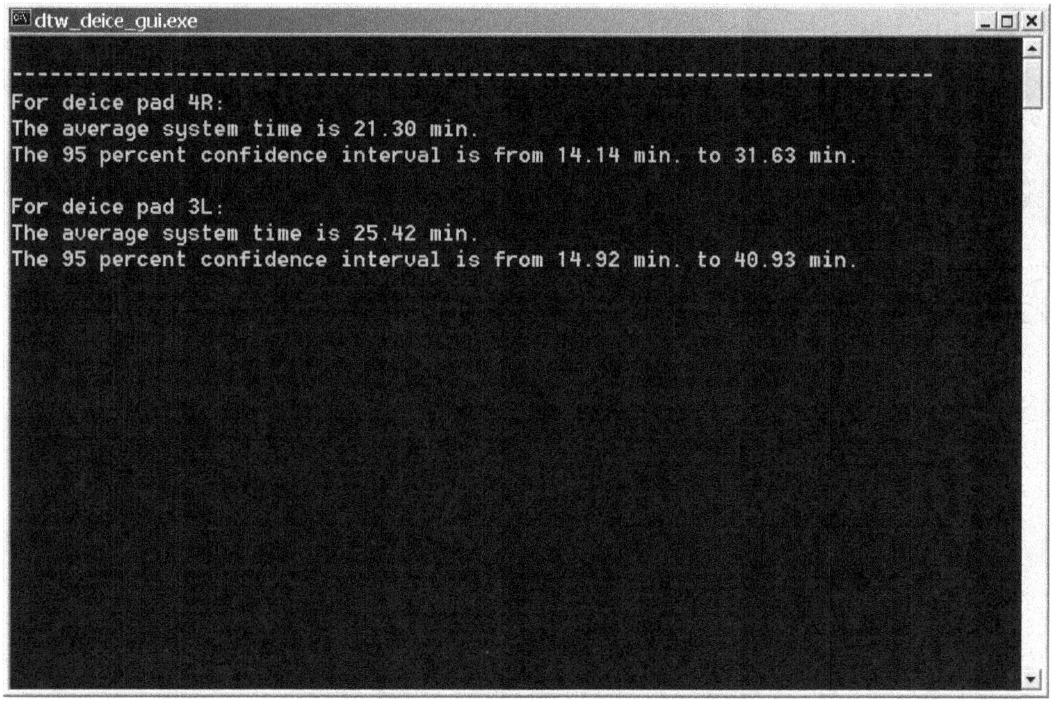

```
For deice pad 4R:
The average system time is 21.30 min.
The 95 percent confidence interval is from 14.14 min. to 31.63 min.

For deice pad 3L:
The average system time is 25.42 min.
The 95 percent confidence interval is from 14.92 min. to 40.93 min.
```

**Figure 9.    Text Display Window**

To obtain estimates of the total time through the deicing process, that is the time in queue plus the time being sprayed, you will need to go through the following steps:

1. Input the status of the deicing positions, either operational or not.
2. Input the aircraft type for each operational deicing position.
3. Input the aircraft type for the queues.
4. Input type of snow. The snow types are: A, B, C, C+2, D, E, and Manual. If necessary, manually input the pad time.
5. Input outbound aircraft type.
6. Click Run Simulation.

While the simulation is running, the word "Simulating" will appear in the results window. Once the simulation is done, the results will be displayed on the result window graphically, as in Figure 8. The two bars in the results window represent the estimate of the statistics for the outbound aircraft to go through each of the deicing pads. The estimated average time is displayed in the middle of the bars. The values above and below the average time represent the upper and lower bounds of the 95 percent confidence interval of the estimated mean.

To exit the tool, click on *File* and select *Exit*.

For more details about each of the steps above, refer to the subsequent subsections.

### 3.3.1 Deicing Pad Status

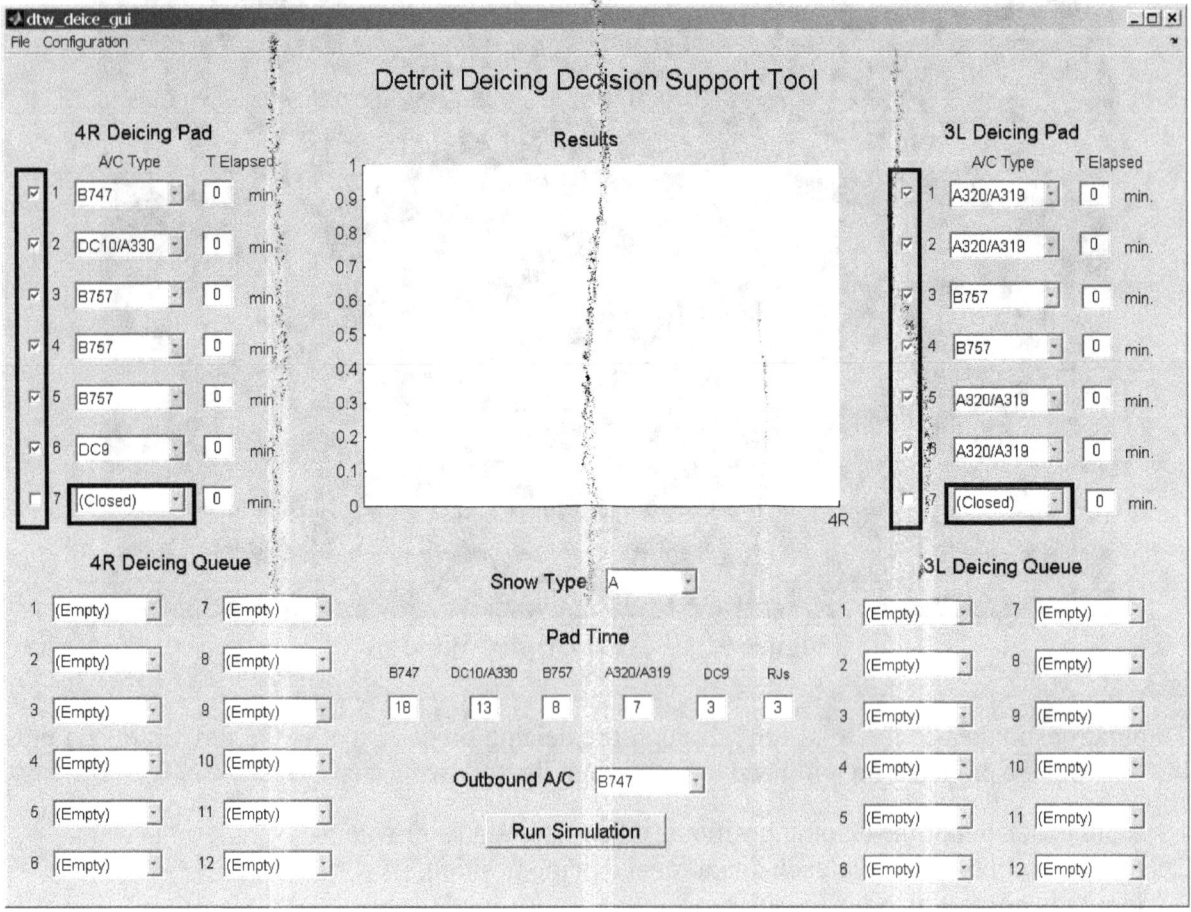

Figure 10.  Deicing Pad Status

The check marks in the rectangular highlight boxes next to the deicing positions in Figure 10 indicate the status of the particular deicing position.

- If the check mark is present, it indicates the deicing position is operational; for example, positions 1-6 in both pads in Figure 10 are operational.
- If the check mark is not present, the corresponding deicing position is not operational (position 7 in both pads in the example). If a position is not operational, the word "Closed" will appear next to the square box.

Change the status of the deicing position by checking or unchecking the boxes next to the particular deicing position.

## 3.3.2 Input Type of Aircraft in the Pad

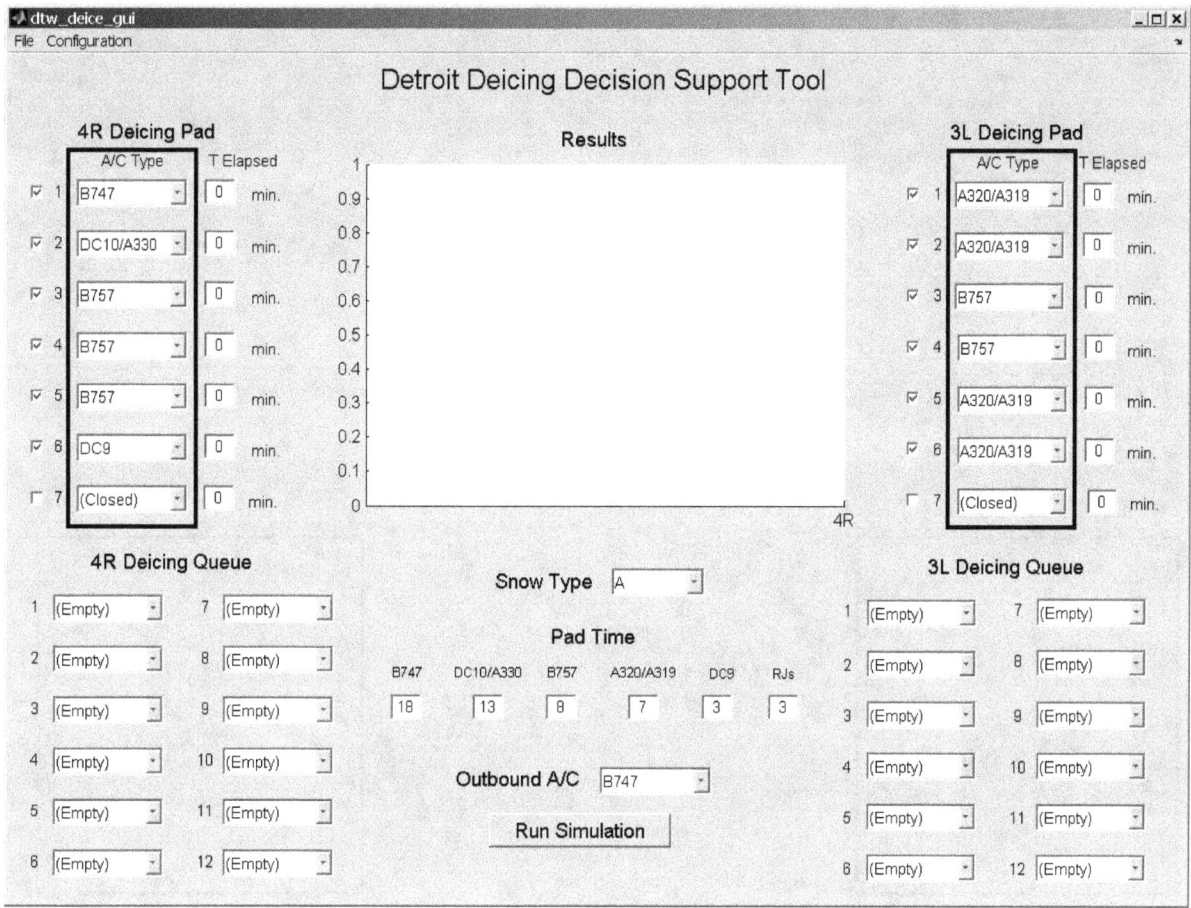

Figure 11. Aircraft Type in the Deicing Pads

Figure 11 highlights the pull-down menus showing the aircraft type for each of the positions. To change the aircraft type,

1. Click on the pull-down menu to the particular deicing position.
2. Select the appropriate aircraft type from the pull-down menu.

To customize the deicing pad configuration, for example, to re-strip the deicing pads, refer to Section 3.4.1.

### 3.3.3 Input Type of Aircraft in the Queue

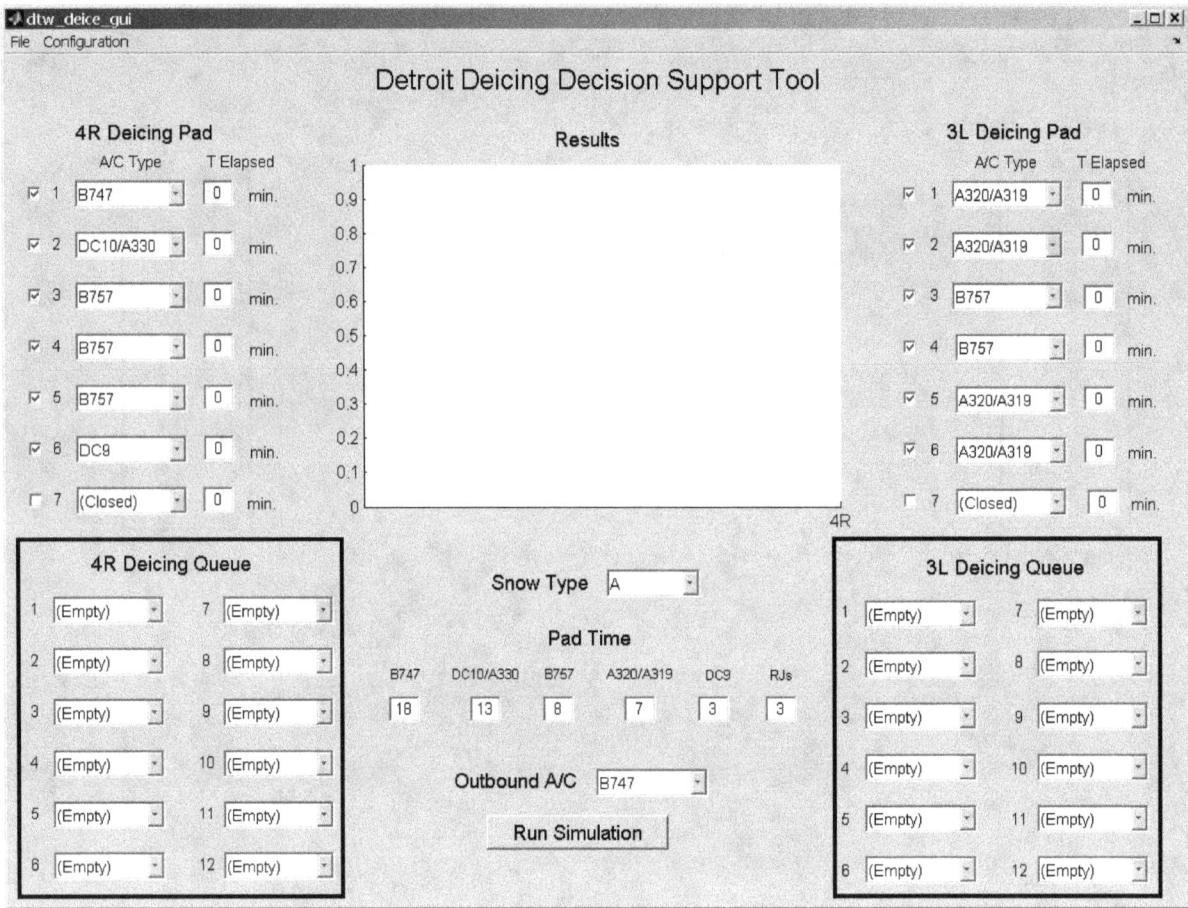

Figure 12.   Aircraft Type in Queue

Figure 12 indicates the areas of the tool to input the aircraft type for the queues. There are twelve possible positions in each of the two queues. To input the aircraft type in the queue:

1. Click on the pull-down menu to the particular deicing position.
2. Select the appropriate aircraft type from the pull-down menu.

To customize the aircraft types in the queues, refer to Section 3.4.1.

### 3.3.4 Input Snow Type

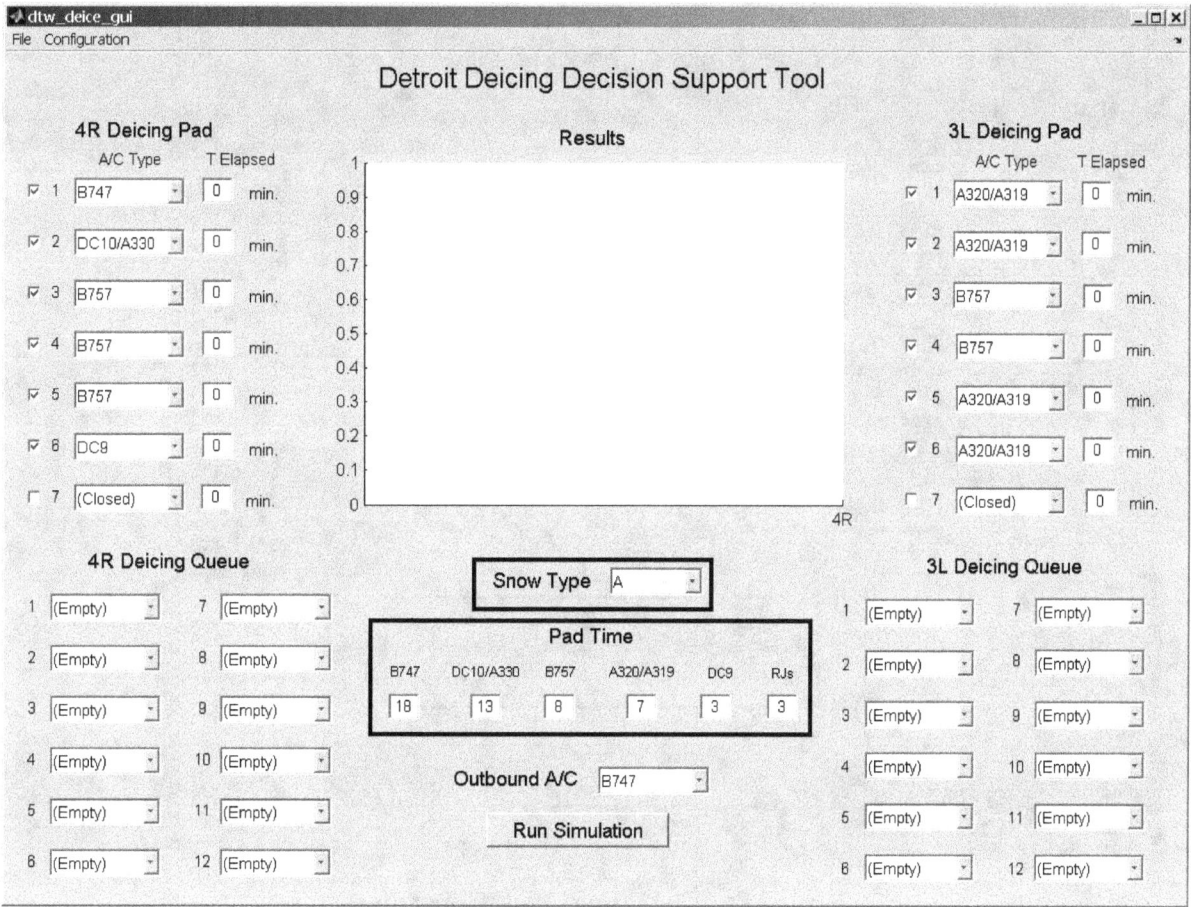

Figure 13.    Snow Type

Figure 13 indicates the areas of the tool to input the snow type, which would dictate the amount of pad time. To input the snow type:

1. Click on the pull-down menu of Snow Type.
2. Select the appropriate snow type.

The pad time for the selected snow type would be displayed in the text boxes below the snow type. There are seven pad times: A, B, C, C + 2 (type C pad time plus two additional minutes), D, E, and Manual (for manual input, see Section 3.3.4.1). For definitions of the different snow types, refer to Northwest Airlines' "Weather Categories Storm Description."

### 3.3.4.1 Manual Input of Pad Time

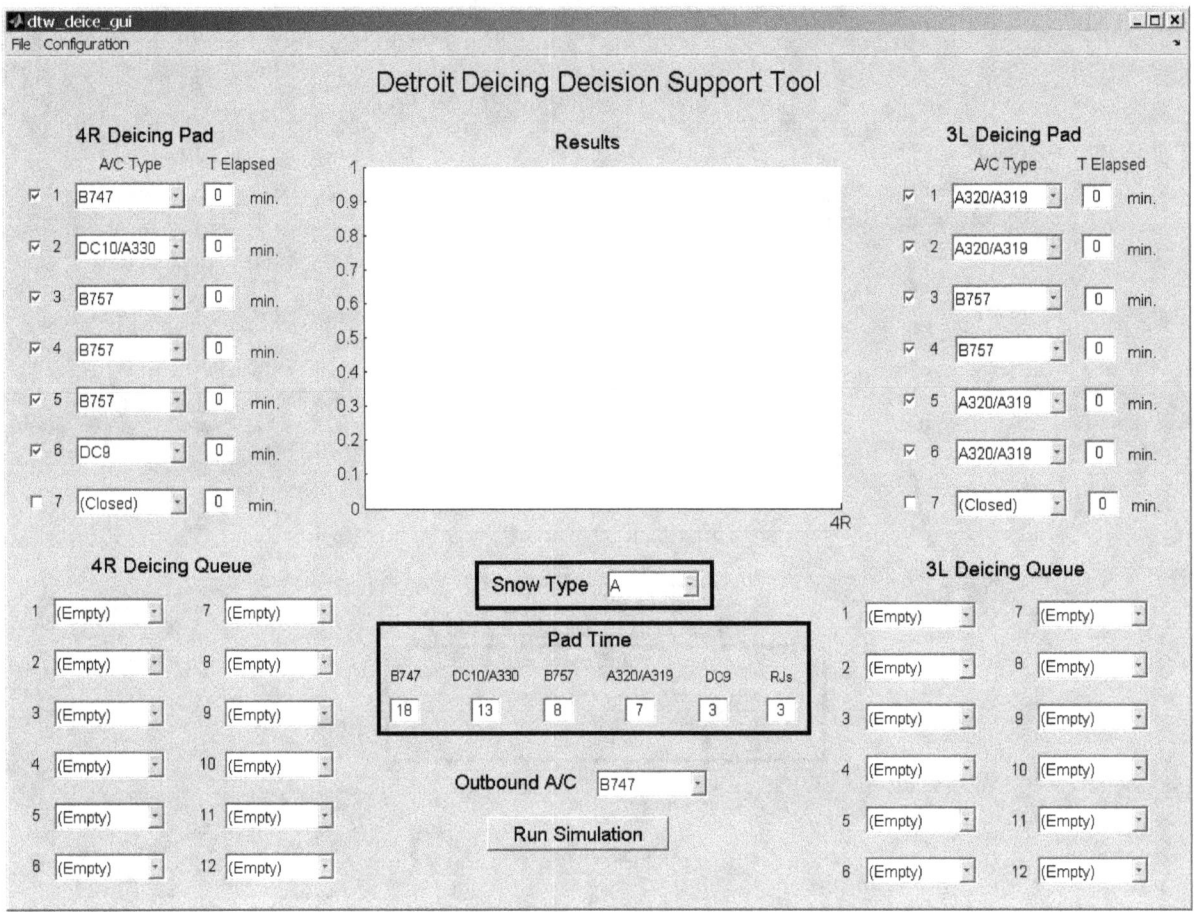

Figure 14.    Manual Pad Time Input

Beside the pad time for Snow Type A, B, C, C+2, D, and E, you can also input the pad time manually. Here is how:

1. In the pull-down menu of Snow Type, select *Manual*, Figure 14.
2. Input the appropriate pad time for each of the aircraft type into the windows below. **Warning**, the values for the pad time must be greater than zero.

To customize some or all of the pad time, refer to Section 3.4.2.

### 3.3.5 Type of Outbound Aircraft

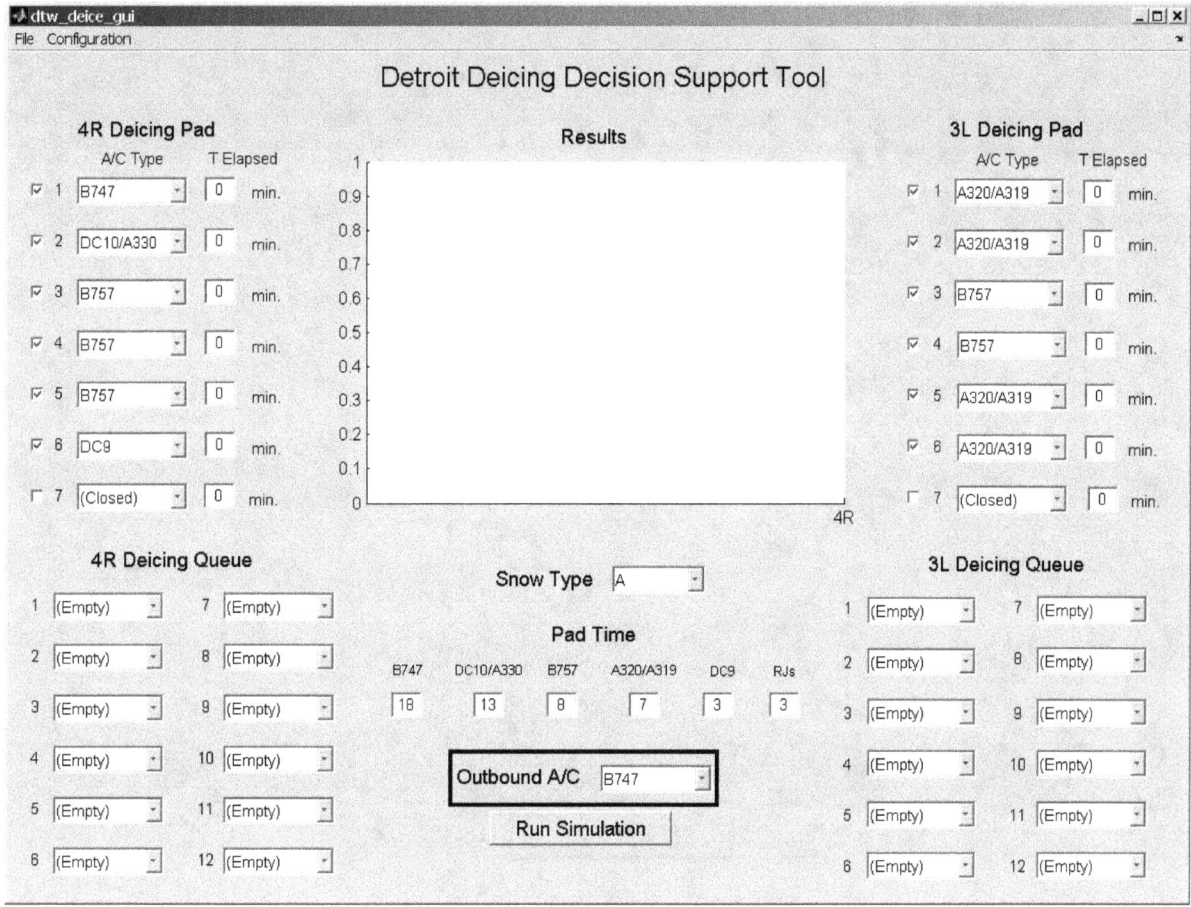

Figure 15. Outbound A/C Type

Figure 15 indicates the areas of the tool to select the outbound aircraft type. To enter the type of outbound aircraft,

1. Click on the outbound aircraft pull-down menu.
2. Select the appropriate aircraft type.

The time estimates from the simulation are the total amounts of time, i.e., waiting in queue and spraying, it would take this outbound A/C to go through either deicing pad 4R or 3L.

### 3.3.6 Run Simulation

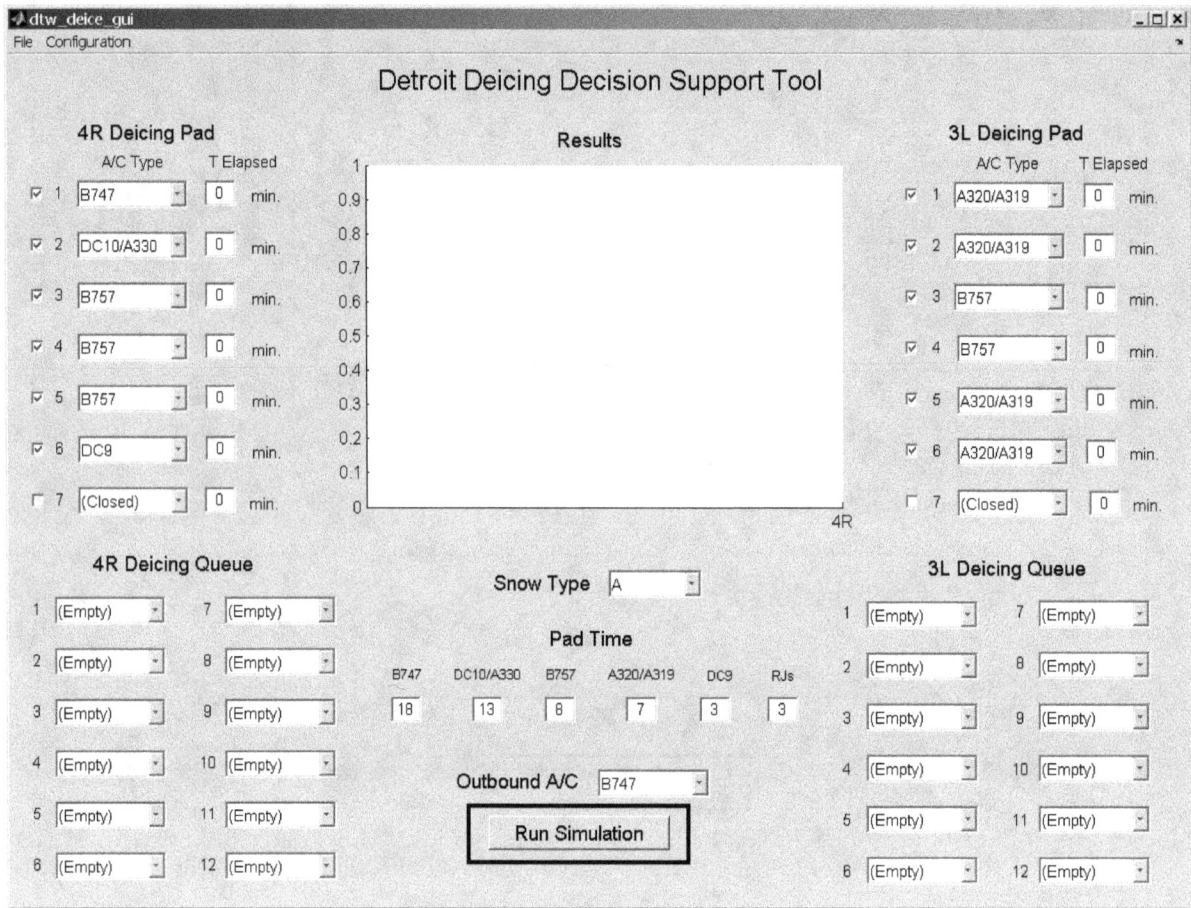

Figure 16.    Start Simulation

After all the necessary inputs are entered, start the simulation by clicking on the *Run Simulation* button, Figure 16. While simulation is running, the results window would display "Simulating." Once the simulation is done, the results will be displayed in the results window.

## 3.4 Customization

Various options can be customized in this section: configuring deicing pads, configuring pad time, and wide-body deicing options.

## 3.4.1 Configure Deicing Pads

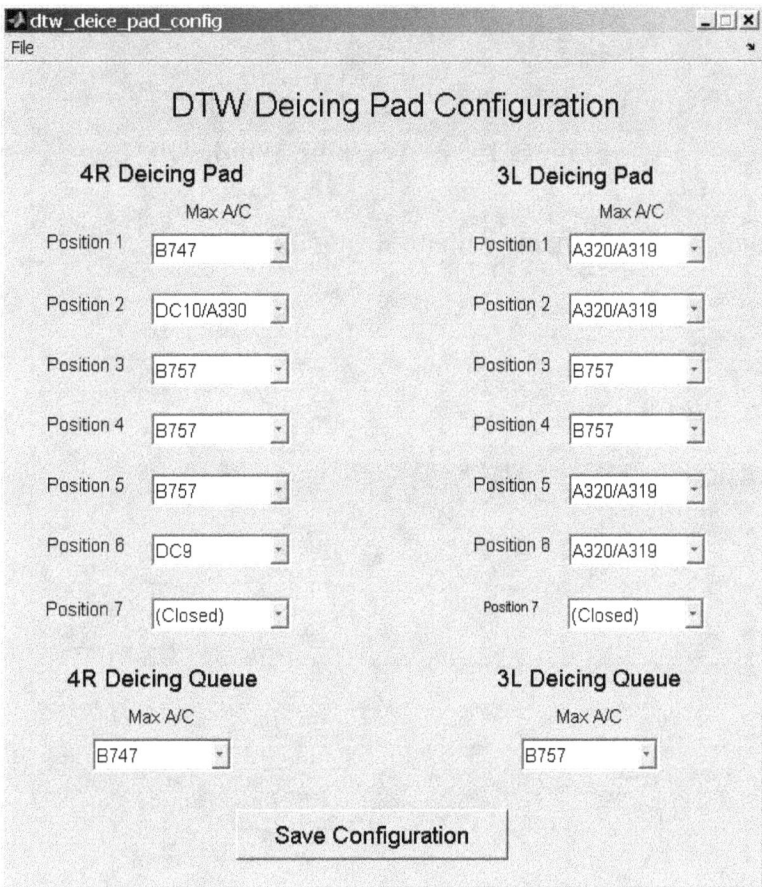

Figure 17.    Pad Configuration Window

To change the configuration of the pads, for example, re-stripping the positions, follow the steps below:

1. Click *Configuration* on top of the tool.
2. Select *Change Pad Configuration*.
3. The Pad Configuration Window will appear, as in Figure 17. The current pad configuration is displayed in the window.
4. Click the pull down menu for each position that needs to be updated. Select the appropriate **maximum aircraft type** for the respective position.
5. Select the **maximum aircraft type** that will be seen in the queue for each pad.
6. When done, click *Save Configuration*.

A warning window, Figure 18, will appear if the maximum aircraft type for the deicing pad is larger or smaller than the maximum A/C type in the queue. Click *OK*. If there is an input error, go through steps 1 thru 6 again. If there is no error, proceed as usual.

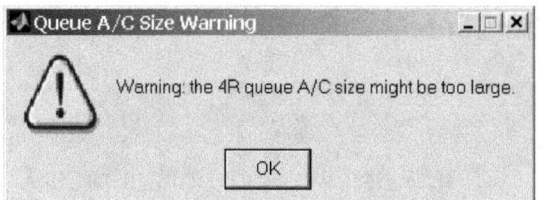

**Figure 18.  Warning Window**

To exit the Pad Configuration Window without saving the changes:

1. Click *File*.
2. Select *Close*.

### 3.4.2  Configure Pad Time

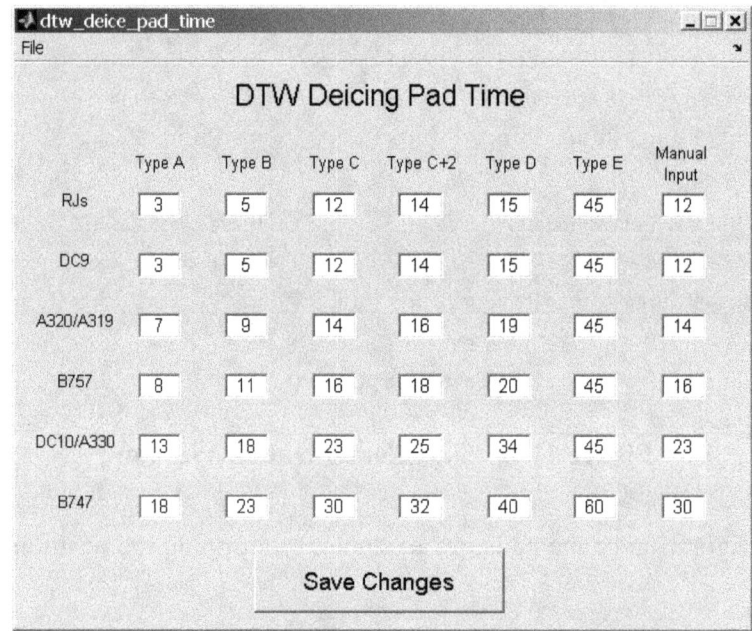

**Figure 19.  Pad Time Configuration Window**

To change the pad time:

1. Click *Configuration* on top of the tool.
2. Select *Change Pad Time*.
3. The Pad Time Configuration Window will appear, as in Figure 19.  The Pad Time Configuration Window displays the current pad time for each aircraft type and each snow type.
4. Select the particular pad time to be changed.
5. Type in the new pad time.  **Warning, the pad time has to be a value greater than zero.**
6. Repeat steps 1 and 2 for each cell that needs to be configured.
7. When done, click *Save Changes* to save the changes and close the Pad Time Configuration Window.

An error window, Figure 20, will appear if any pad time is not greater than zero. To correct the error, click *OK*, then go through steps 1 thru 7 again.

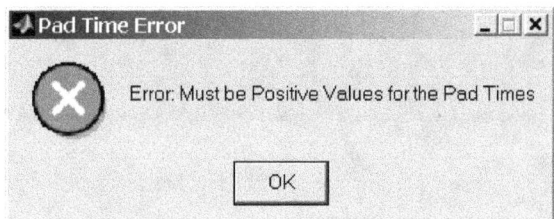

**Figure 20.    Error Window**

To exit the Pad Time Configuration Window without saving the changes:

1. Click *File*.
2. Select *Close*.

### 3.4.3 Wide-Body Deicing

The deicing DST allows for wide-body deicing using two adjacent narrow-body positions when they are open. This feature only applies to deicing pad 4R. To enable or disable this feature, follow the steps below:

1. Select *Configuration* on the top of the tool.
2. Check or uncheck Wide-Body Deicing in Narrow-Body Positions.

If *Wide-Body Deicing in Narrow-Body Positions* is checked, it means the wide-body can be deiced in two adjacent narrow-body positions. If it is not checked, this feature is disabled.

## 3.5 Disclaimer

The developer of the deicing DST is not responsible for any computer system damage that occurs during the installation or operation of the Detroit Deicing Decision Support Tool.

# 4. Results

In this section, some simulation results will be presented. In each of the scenarios, the types of aircraft that are already in the deicing positions are specified. In addition, the types of aircraft in the queues to each of the deicing pads are declared. The user also needs to enter the snow type as well as the aircraft of interest. The output of the simulation is the estimates of total system times associated with the outbound aircraft deiced in either one of the two deicing pads.

## 4.1 Snow Type

In this simulation, the impact of different snow types on the simulation will be shown. To enable equal comparison between deicing at the two pads, we set the type of aircraft in the corresponding deicing positions to be identical for both deicing pads, while satisfying the size constraint, to enable equal comparison between deicing at the two pads, i.e., A320 in Positions 1, 2, and 5; B757 in Positions 3 and 4; and DC-9 in Position 6. There is no aircraft in queue and the outbound aircraft is an A320. We entered two types of snow events, A and C. The total system times are show in Figure 21 and Figure 22. As the type of snow changed from a less severe weather pattern (A) to a more severe pattern (C), the corresponding estimate of the total system time also increased due to a lengthier deicing process.

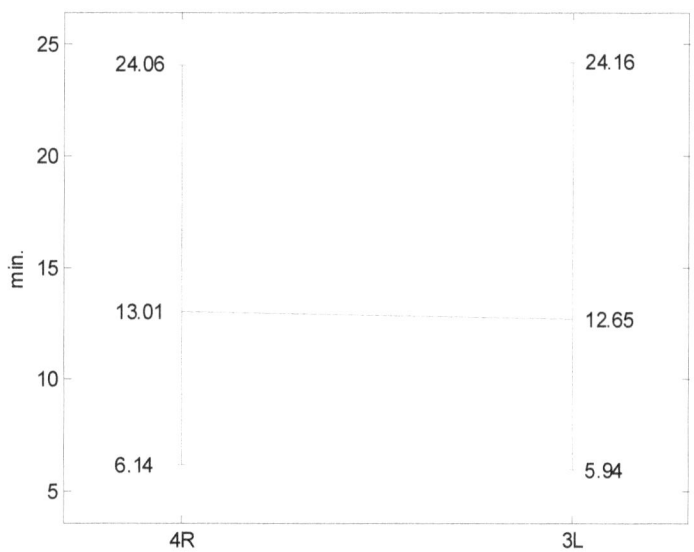

**Figure 21.     Type A Snow**

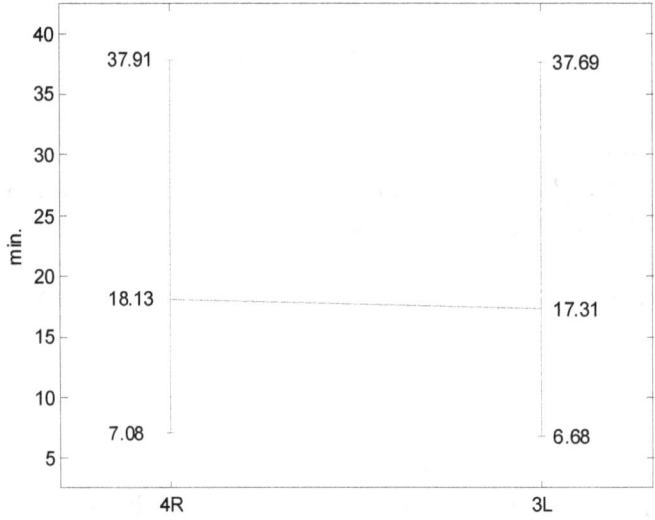

**Figure 22.    Type C Snow, Outbound A320**

In this scenario, it may be advantageous to send the outbound A320 to deicing pad 3L regardless of the type of snow, since the estimated total system times are shorter for deicing pad 3L in both snow types. The slight difference in the estimate of the total system time between deicing pad 4R and 3L is due to the differences in pad configuration. In 4R, out of the six positions, only Position 1 through 5 can accommodate an A320, whereas all six positions at deicing pad 3L can accommodate an A320. The shortage of one deicing position in deicing pad 4L contributed to the slightly lower total system time.

## 4.2  Aircraft Type

Here, we show the impact of different aircraft types to the estimate total system time. We set the type of aircraft in the deicing positions to be identical for both deicing pads, as in scenario above. There is no aircraft in queue with type C snow. The outbound aircraft is either an A320 or B757. The result for the A320 is shown in Figure 22 and the result for B757 is shown in Figure 23.

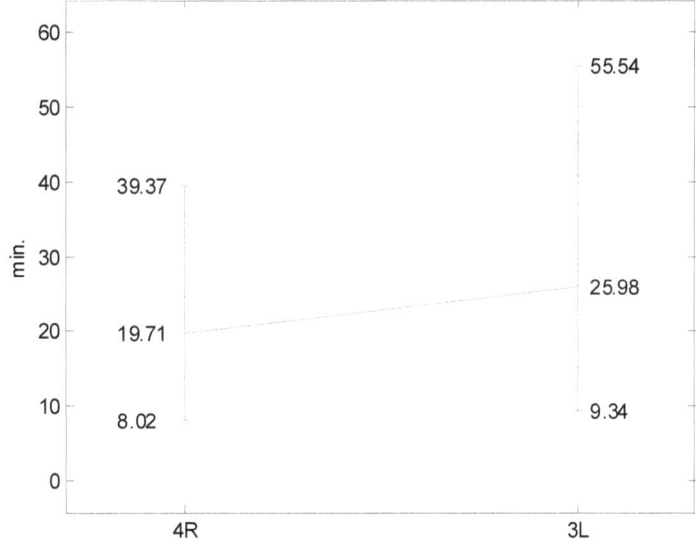

**Figure 23.   Outbound B757**

No one deicing pad consistently provided a shorter estimated total system time for both outbound aircraft types. In the case of an outbound A320, it is better to send it to deicing pad 3L, as explained in the previous section. On the other hand, an outbound B757 would be more efficient if it is sent to deicing pad 4R. Once again, this is due to the different physical layouts of the deicing pads. Position 1 through 5 of deicing pad 4R can accommodate a B757 and only Position 3 and 4 of deicing pad 3L can deice a B757. Therefore, this results in a much shorter estimated total system time for the outbound 757 using deicing pad 4R rather than deicing pad 3L.

## 4.3 Nonlinearity

In this section, we tested the effect of the queue order on the estimated total system time. Once again, we set the type of aircraft in the deicing positions to be identical for both deicing pads, as in Scenarios 2 and 3. The snow type is C. The queues for both deicing pads 4R and 3L are also set to be the same. In the first queue, there are two B757s in the first two positions of the queue, followed by two A320s. In the second queue, the order is reversed, i.e., there are two A320s in the front of the queue, followed by two B757s. The outbound aircraft is a B757. The corresponding estimated total system times are shown in Figure 24 for the first queue and in Figure 25 for the second queue.

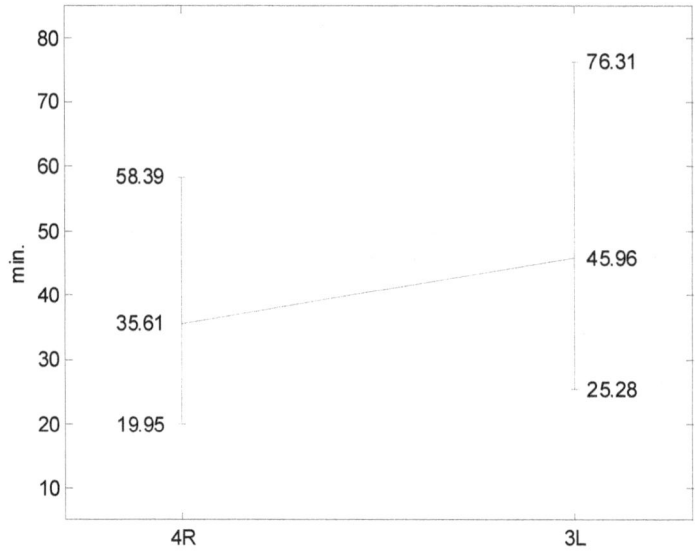

**Figure 24.    Two B757s at Front of Queue**

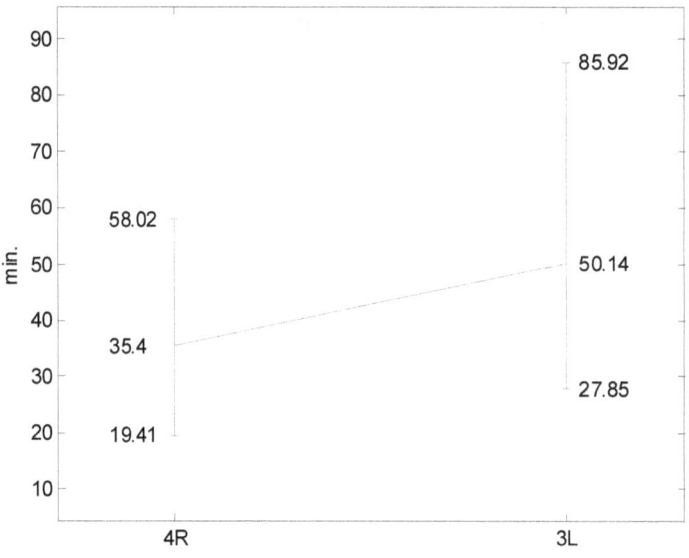

**Figure 25.    Two A320s at Front of Queue**

The two different queue orders would result in similar total system times if the outbound B757 goes through deicing pad 4R. This is due to the five positions at deicing pad 4R, which can accommodate a B757. On the other hand, there is a noticeable difference in the total system times associated with the two queue orders if the outbound B757 is to go through deicing pad 3L. In the first case, the two B757s in the front of the queue would hold up all the aircraft behind it while Position 3 and 4 are not available in deicing pad 3L. During the wait, the smaller deicing positions may become free. Once the B757s are able to be deiced, there is a higher chance that the A320s can be deiced, leaving the outbound B757 with a shorter wait. In the second case, there is some probability that either or both of Position 3 and 4 would become free and the A320s in the front of the queue could take either opening. Therefore, the two B757s in the back of the queue would be held up while holding back the outbound B757, which is added to the end of the queue. Since there are only two B757-capable deicing positions, the two B757s in the back of the queue would take those positions and hold the outbound B757 at the queue.

# 5. Future Directions

The usefulness of the deicing tool can be further enhanced by populating the various inputs with real-time surface surveillance. With real-time inputs, the pad time can be updated continuously to more accurately reflect variability between the deicing positions. This would enable better estimates of total system time. Other enhancements include scheduled shut down of deicing positions in the middle of the simulation, to account for additional variability. Most of these augmentations improve the tactical decision making. This tool can also be used for strategic planning if future flight schedules are known.

# 6. Contact Information

For further information about the deicing decision support tool or the installation CD-ROM, please contact Jonathan Lee at (617) 494-2411 or <jonathan.lee@volpe.dot.gov>, Volpe National Transportation Systems Center, U.S. Department of Transportation, Cambridge, MA 02142.

# References

1. Beatty, R., R. Hsu, L. Berry, and J. Rome. "Preliminary Evaluation of Flight Delay Propagation through an Airline Schedule," 2$^{nd}$ USA/Europe Air Traffic Management R&D Seminar, Orlando, FL, Dec. 1-4, 1998.

2. Cassandras, Christos. *Discrete Event System: Modeling and Performance Analysis*. Irwin and Aksen Associates, 1993.

3. Federal Aviation Administration. "AND-500 Performance Metrics Results to Date," Oct. 2003.

4. Mueller, Eric R., and G. B. Chatterji. "Analysis of Aircraft Arrival and Departure Delay Characteristics," presented at AIAA 2$^{nd}$ Aircraft Technology, Integration, and Operations Forum, Los Angeles, CA, Oct. 1-3, 2002, AIAA 2002-5866.

5. Northwest Airlines. "Weather Categories," internal document.

## Appendix A. Volpe Project Team

The Volpe team working on the Detroit deicing project is composed of Michael Geyer, Anastasios Daskalakis, Jonathan Lee, and Suzanne Chen.

# Appendix B. Detroit Deicing Decision Support Tool, Build 2, MATLAB Code

```
1 function varargout = dtw_deice_gui(varargin)
2 % DTW_DEICE_GUI M-file for dtw_deice_gui.fig
3 % DTW_DEICE_GUI, by itself, creates a new DTW_DEICE_GUI or raises the existing
4 % singleton*.
5 %
6 % H = DTW_DEICE_GUI returns the handle to a new DTW_DEICE_GUI or the handle to
7 % the existing singleton*.
8 %
9 % DTW_DEICE_GUI('CALLBACK',hObject,eventData,handles,...) calls the local
10 % function named CALLBACK in DTW_DEICE_GUI.M with the given input arguments.
11 %
12 % DTW_DEICE_GUI('Property','Value',...) creates a new DTW_DEICE_GUI or raises the
13 % existing singleton*. Starting from the left, property value pairs are
14 % applied to the GUI before dtw_deice_gui_OpeningFunction gets called. An
15 % unrecognized property name or invalid value makes property application
16 % stop. All inputs are passed to dtw_deice_gui_OpeningFcn via varargin.
17 %
18 % *See GUI Options on GUIDE's Tools menu. Choose "GUI allows only one
19 % instance to run (singleton)".
20 %
21 % See also: GUIDE, GUIDATA, GUIHANDLES
22
23 % Edit the above text to modify the response to help dtw_deice_gui
24
25 % Last Modified by GUIDE v2.5 26-Aug-2005 17:36:41
26
27 % Begin initialization code - DO NOT EDIT
28 gui_Singleton = 1;
29 gui_State = struct('gui_Name', mfilename, ...
30 'gui_Singleton', gui_Singleton, ...
31 'gui_OpeningFcn', @dtw_deice_gui_OpeningFcn, ...
32 'gui_OutputFcn', @dtw_deice_gui_OutputFcn, ...
33 'gui_LayoutFcn', [] , ...
34 'gui_Callback', []);
35 if nargin & isstr(varargin{1})
36 gui_State.gui_Callback = str2func(varargin{1});
37 end
38
39 if nargout
40 [varargout{1:nargout}] = gui_mainfcn(gui_State, varargin{:});
41 else
42 gui_mainfcn(gui_State, varargin{:});
43 end
44 % End initialization code - DO NOT EDIT
45
46
47 % --- Executes just before dtw_deice_gui is made visible.
48 function dtw_deice_gui_OpeningFcn(hObject, eventdata, handles, varargin)
49 % This function has no output args, see OutputFcn.
50 % hObject handle to figure
51 % eventdata reserved - to be defined in a future version of MATLAB
52 % handles structure with handles and user data (see GUIDATA)
53 % varargin command line arguments to dtw_deice_gui (see VARARGIN)
54
55 % Choose default command line output for dtw_deice_gui
```

```matlab
56 handles.output = hObject;
57
58 % Update handles structure
59 guidata(hObject, handles);
60
61 % UIWAIT makes dtw_deice_gui wait for user response (see UIRESUME)
62 % uiwait(handles.dtw_deice_gui);
63
64 % Center the GUI
65 movegui(handles.dtw_deice_gui, 'center');
66
67 % --- Outputs from this function are returned to the command line.
68 function varargout = dtw_deice_gui_OutputFcn(hObject, eventdata, handles)
69 % varargout cell array for returning output args (see VARARGOUT);
70 % hObject handle to figure
71 % eventdata reserved - to be defined in a future version of MATLAB
72 % handles structure with handles and user data (see GUIDATA)
73
74 % Get default command line output from handles structure
75 varargout{1} = handles.output;
76
77
78
79 % -------------------------------------------------------------------------
80 % Load values for GUI objects
81 % -------------------------------------------------------------------------
82
83 % Load configuration data
84 load dtw_deice_data.mat;
85
86 % Set configuration for deicing pad 4R
87 for count = 1:7,
88 eval(['set(handles.type_s_4_', num2str(count), ', ''String'', ac_list_s_4_', num2str(count), ');']);
89 if (eval(['length(ac_list_s_4_', num2str(count), ') > 1']))
90 eval(['set(handles.checkbox_4_', num2str(count), ', ''Value'', 1);']);
91 else
92 eval(['set(handles.checkbox_4_', num2str(count), ', ''Value'', 0);']);
93 end
94 end
95
96 % Set configuration for the queue for deicing pad 4R
97 for count = 1:12,
98 eval(['set(handles.type_q_4_', num2str(count), ', ''String'', ac_list_q_4);']);
99 end
100
101 % Set configuration for deicing pad 3L
102 for count = 1:7,
103 eval(['set(handles.type_s_3_', num2str(count), ', ''String'', ac_list_s_3_', num2str(count), ');']);
104 if (eval(['length(ac_list_s_3_', num2str(count), ') ~= 1']))
105 eval(['set(handles.checkbox_3_', num2str(count), ', ''Value'', 1);']);
106 else
107 eval(['set(handles.checkbox_3_', num2str(count), ', ''Value'', 0);']);
108 end
109 end
110
111 % Set configuration for the queue for deicing pad 3L
112 for count = 1:12,
113 eval(['set(handles.type_q_3_', num2str(count), ', ''String'', ac_list_q_3);']);
```

```matlab
114 end
115
116 % Set snow type and pad time
117 set(handles.type_snow, 'Value', 1);
118 set(handles.time_rj, 'String', num2str(pad_time(1,1)));
119 set(handles.time_dc9, 'String', num2str(pad_time(2,1)));
120 set(handles.time_a320, 'String', num2str(pad_time(3,1)));
121 set(handles.time_b757, 'String', num2str(pad_time(4,1)));
122 set(handles.time_dc10, 'String', num2str(pad_time(5,1)));
123 set(handles.time_b747, 'String', num2str(pad_time(6,1)));
124
125 % Set weather wide-body can or cannot be deiced in two adjacent
126 % narrow-body positions
127 if widebody_flag == 0,
128 set(handles.menu_widebody, 'Checked', 'off');
129 elseif widebody_flag == 1,
130 set(handles.menu_widebody, 'Checked', 'on');
131 end
132
133 %-------------------------------------------------------------------------
134 % Creat GUI objects
135 %-------------------------------------------------------------------------
136
137 % -----------------------------------------------------------------------
138 function menu_file_Callback(hObject, eventdata, handles)
139 % hObject handle to menu_file (see GCBO)
140 % eventdata reserved - to be defined in a future version of MATLAB
141 % handles structure with handles and user data (see GUIDATA)
142
143
144 % -----------------------------------------------------------------------
145 function menu_file_close_Callback(hObject, eventdata, handles)
146 % hObject handle to Untitled_2 (see GCBO)
147 % eventdata reserved - to be defined in a future version of MATLAB
148 % handles structure with handles and user data (see GUIDATA)
149 close(handles.dtw_deice_gui);
150
151
152 % -----------------------------------------------------------------------
153 function menu_pad_config_Callback(hObject, eventdata, handles)
154 % hObject handle to menu_pad_config (see GCBO)
155 % eventdata reserved - to be defined in a future version of MATLAB
156 % handles structure with handles and user data (see GUIDATA)
157
158
159 % -----------------------------------------------------------------------
160 function menu_change_pad_config_Callback(hObject, eventdata, handles)
161 % hObject handle to menu_change_pad_config (see GCBO)
162 % eventdata reserved - to be defined in a future version of MATLAB
163 % handles structure with handles and user data (see GUIDATA)
164
165 % Change the pad configuration subroutine.
166 dtw_deice_pad_config;
167
168
169 % -----------------------------------------------------------------------
170 function menu_change_pad_time_Callback(hObject, eventdata, handles)
171 % hObject handle to menu_change_pad_time (see GCBO)
172 % eventdata reserved - to be defined in a future version of MATLAB
173 % handles structure with handles and user data (see GUIDATA)
174
175 % Change the pad time
```

```
176 dtw_deice_pad_time;
177
178
179 % -----------------------------------------------------------------
180 function menu_widebody_Callback(hObject, eventdata, handles)
181 % hObject handle to menu_widebody (see GCBO)
182 % eventdata reserved - to be defined in a future version of MATLAB
183 % handles structure with handles and user data (see GUIDATA)
184
185 load dtw_deice_data.mat;
186
187 % Update the widebody_flag value
188 widebody_flag = mod(widebody_flag + 1, 2);
189 % Set weather wide-body can or cannot be deiced in two adjucent
190 % narrow-body positions
191 if widebody_flag == 0,
192 set(handles.menu_widebody, 'Checked', 'off');
193 elseif widebody_flag == 1,
194 set(handles.menu_widebody, 'Checked', 'on');
195 end
196
197 % Save the variables
198 save dtw_deice_data.mat ac_list_s* ac_list_q* pad_time widebody_flag;
199
200 % --- Executes during object creation, after setting all properties.
201 function type_snow_CreateFcn(hObject, eventdata, handles)
202 % hObject handle to type_snow (see GCBO)
203 % eventdata reserved - to be defined in a future version of MATLAB
204 % handles empty - handles not created until after all CreateFcns called
205
206 % Hint: popupmenu controls usually have a white background on Windows.
207 %   See ISPC and COMPUTER.
208 if ispc
209 set(hObject,'BackgroundColor','white');
210 else
211 set(hObject,'BackgroundColor',get(0,'defaultUicontrolBackgroundColor'));
212 end
213
214 % --- Executes on selection change in type_snow.
215 function type_snow_Callback(hObject, eventdata, handles)
216 % hObject handle to type_snow (see GCBO)
217 % eventdata reserved - to be defined in a future version of MATLAB
218 % handles structure with handles and user data (see GUIDATA)
219
220 % Hints: contents = get(hObject,'String') returns type_snow contents as cell array
221 %        contents{get(hObject,'Value')} returns selected item from
222 %        type_snow
223
224 % Refresh the snow time
225 snow_type = get(handles.type_snow, 'Value');
226
227 % Load pad time
228 load dtw_deice_data;
229
230 % Set pad time
231 set(handles.time_rj, 'String', num2str(pad_time(1, snow_type)));
232 set(handles.time_dc9, 'String', num2str(pad_time(2, snow_type)));
233 set(handles.time_a320, 'String', num2str(pad_time(3, snow_type)));
234 set(handles.time_b757, 'String', num2str(pad_time(4, snow_type)));
235 set(handles.time_dc10, 'String', num2str(pad_time(5, snow_type)));
236 set(handles.time_b747, 'String', num2str(pad_time(6, snow_type)));
237
```

```matlab
% --- Executes during object creation, after setting all properties.
function time_b747_CreateFcn(hObject, eventdata, handles)
% hObject    handle to time_b747 (see GCBO)
% eventdata  reserved - to be defined in a future version of MATLAB
% handles    empty - handles not created until after all CreateFcns called

% Hint: edit controls usually have a white background on Windows.
%       See ISPC and COMPUTER.
if ispc
set(hObject,'BackgroundColor','white');
else
set(hObject,'BackgroundColor',get(0,'defaultUicontrolBackgroundColor'));
end

function time_b747_Callback(hObject, eventdata, handles)
% hObject    handle to time_b747 (see GCBO)
% eventdata  reserved - to be defined in a future version of MATLAB
% handles    structure with handles and user data (see GUIDATA)

% Hints: get(hObject,'String') returns contents of time_b747 as text
%        str2double(get(hObject,'String')) returns contents of time_b747 as a double


% --- Executes during object creation, after setting all properties.
function time_dc10_CreateFcn(hObject, eventdata, handles)
% hObject    handle to time_dc10 (see GCBO)
% eventdata  reserved - to be defined in a future version of MATLAB
% handles    empty - handles not created until after all CreateFcns called

% Hint: edit controls usually have a white background on Windows.
%       See ISPC and COMPUTER.
if ispc
set(hObject,'BackgroundColor','white');
else
set(hObject,'BackgroundColor',get(0,'defaultUicontrolBackgroundColor'));
end

function time_dc10_Callback(hObject, eventdata, handles)
% hObject    handle to time_dc10 (see GCBO)
% eventdata  reserved - to be defined in a future version of MATLAB
% handles    structure with handles and user data (see GUIDATA)

% Hints: get(hObject,'String') returns contents of time_dc10 as text
%        str2double(get(hObject,'String')) returns contents of time_dc10 as a double


% --- Executes during object creation, after setting all properties.
function time_b757_CreateFcn(hObject, eventdata, handles)
% hObject    handle to time_b757 (see GCBO)
% eventdata  reserved - to be defined in a future version of MATLAB
% handles    empty - handles not created until after all CreateFcns called

% Hint: edit controls usually have a white background on Windows.
%       See ISPC and COMPUTER.
if ispc
set(hObject,'BackgroundColor','white');
else
set(hObject,'BackgroundColor',get(0,'defaultUicontrolBackgroundColor'));
end

function time_b757_Callback(hObject, eventdata, handles)
```

```
300 % hObject handle to time_b757 (see GCBO)
301 % eventdata reserved - to be defined in a future version of MATLAB
302 % handles structure with handles and user data (see GUIDATA)
303
304 % Hints: get(hObject,'String') returns contents of time_b757 as text
305 % str2double(get(hObject,'String')) returns contents of time_b757 as a double
306
307
308 % --- Executes during object creation, after setting all properties.
309 function time_a320_CreateFcn(hObject, eventdata, handles)
310 % hObject handle to time_a320 (see GCBO)
311 % eventdata reserved - to be defined in a future version of MATLAB
312 % handles empty - handles not created until after all CreateFcns called
313
314 % Hint: edit controls usually have a white background on Windows.
315 % See ISPC and COMPUTER.
316 if ispc
317 set(hObject,'BackgroundColor','white');
318 else
319 set(hObject,'BackgroundColor',get(0,'defaultUicontrolBackgroundColor'));
320 end
321
322 function time_a320_Callback(hObject, eventdata, handles)
323 % hObject handle to time_a320 (see GCBO)
324 % eventdata reserved - to be defined in a future version of MATLAB
325 % handles structure with handles and user data (see GUIDATA)
326
327 % Hints: get(hObject,'String') returns contents of time_a320 as text
328 % str2double(get(hObject,'String')) returns contents of time_a320 as a double
329
330
331 % --- Executes during object creation, after setting all properties.
332 function time_dc9_CreateFcn(hObject, eventdata, handles)
333 % hObject handle to time_dc9 (see GCBO)
334 % eventdata reserved - to be defined in a future version of MATLAB
335 % handles empty - handles not created until after all CreateFcns called
336
337 % Hint: edit controls usually have a white background on Windows.
338 % See ISPC and COMPUTER.
339 if ispc
340 set(hObject,'BackgroundColor','white');
341 else
342 set(hObject,'BackgroundColor',get(0,'defaultUicontrolBackgroundColor'));
343 end
344
345 function time_dc9_Callback(hObject, eventdata, handles)
346 % hObject handle to time_dc9 (see GCBO)
347 % eventdata reserved - to be defined in a future version of MATLAB
348 % handles structure with handles and user data (see GUIDATA)
349
350 % Hints: get(hObject,'String') returns contents of time_dc9 as text
351 % str2double(get(hObject,'String')) returns contents of time_dc9 as a double
352
353
354 % --- Executes during object creation, after setting all properties.
355 function time_rj_CreateFcn(hObject, eventdata, handles)
356 % hObject handle to time_rj (see GCBO)
357 % eventdata reserved - to be defined in a future version of MATLAB
358 % handles empty - handles not created until after all CreateFcns called
359
360 % Hint: edit controls usually have a white background on Windows.
```

```matlab
% See ISPC and COMPUTER.
if ispc
    set(hObject,'BackgroundColor','white');
else
    set(hObject,'BackgroundColor',get(0,'defaultUicontrolBackgroundColor'));
end

function time_rj_Callback(hObject, eventdata, handles)
% hObject handle to time_rj (see GCBO)
% eventdata reserved - to be defined in a future version of MATLAB
% handles structure with handles and user data (see GUIDATA)

% Hints: get(hObject,'String') returns contents of time_rj as text
%        str2double(get(hObject,'String')) returns contents of time_rj as a double


% --- Executes on button press in checkbox_4_1.
function checkbox_4_1_Callback(hObject, eventdata, handles)
% hObject handle to checkbox_4_1 (see GCBO)
% eventdata reserved - to be defined in a future version of MATLAB
% handles structure with handles and user data (see GUIDATA)

% Hint: get(hObject,'Value') returns toggle state of checkbox_4_1
load dtw_deice_data.mat;
if (get(handles.checkbox_4_1, 'Value') == 1)
set(handles.type_s_4_1, 'String', ac_list_s_4_1);
elseif (get(handles.checkbox_4_1, 'Value') == 0)
set(handles.type_s_4_1, 'String', {'(Closed)'});
end


% --- Executes on button press in checkbox_4_2.
function checkbox_4_2_Callback(hObject, eventdata, handles)
% hObject handle to checkbox_4_2 (see GCBO)
% eventdata reserved - to be defined in a future version of MATLAB
% handles structure with handles and user data (see GUIDATA)

% Hint: get(hObject,'Value') returns toggle state of checkbox_4_2
load dtw_deice_data.mat;
if (get(handles.checkbox_4_2, 'Value') == 1)
set(handles.type_s_4_2, 'String', ac_list_s_4_2);
elseif (get(handles.checkbox_4_2, 'Value') == 0)
set(handles.type_s_4_2, 'String', {'(Closed)'});
end


% --- Executes on button press in checkbox_4_3.
function checkbox_4_3_Callback(hObject, eventdata, handles)
% hObject handle to checkbox_4_3 (see GCBO)
% eventdata reserved - to be defined in a future version of MATLAB
% handles structure with handles and user data (see GUIDATA)

% Hint: get(hObject,'Value') returns toggle state of checkbox_4_3
load dtw_deice_data.mat;
if (get(handles.checkbox_4_3, 'Value') == 1)
set(handles.type_s_4_3, 'String', ac_list_s_4_3);
elseif (get(handles.checkbox_4_3, 'Value') == 0)
set(handles.type_s_4_3, 'String', {'(Closed)'});
end


% --- Executes on button press in checkbox_4_4.
function checkbox_4_4_Callback(hObject, eventdata, handles)
```

```matlab
% hObject    handle to checkbox_4_4 (see GCBO)
% eventdata  reserved - to be defined in a future version of MATLAB
% handles    structure with handles and user data (see GUIDATA)

% Hint: get(hObject,'Value') returns toggle state of checkbox_4_4
load dtw_deice_data.mat;
if (get(handles.checkbox_4_4, 'Value') == 1)
    set(handles.type_s_4_4, 'String', ac_list_s_4_4);
elseif (get(handles.checkbox_4_4, 'Value') == 0)
    set(handles.type_s_4_4, 'String', {'(Closed)'});
end


% --- Executes on button press in checkbox_4_5.
function checkbox_4_5_Callback(hObject, eventdata, handles)
% hObject    handle to checkbox_4_5 (see GCBO)
% eventdata  reserved - to be defined in a future version of MATLAB
% handles    structure with handles and user data (see GUIDATA)

% Hint: get(hObject,'Value') returns toggle state of checkbox_4_5
load dtw_deice_data.mat;
if (get(handles.checkbox_4_5, 'Value') == 1)
    set(handles.type_s_4_5, 'String', ac_list_s_4_5);
elseif (get(handles.checkbox_4_5, 'Value') == 0)
    set(handles.type_s_4_5, 'String', {'(Closed)'});
end


% --- Executes on button press in checkbox_4_6.
function checkbox_4_6_Callback(hObject, eventdata, handles)
% hObject    handle to checkbox_4_6 (see GCBO)
% eventdata  reserved - to be defined in a future version of MATLAB
% handles    structure with handles and user data (see GUIDATA)

% Hint: get(hObject,'Value') returns toggle state of checkbox_4_6
load dtw_deice_data.mat;
if (get(handles.checkbox_4_6, 'Value') == 1)
    set(handles.type_s_4_6, 'String', ac_list_s_4_6);
elseif (get(handles.checkbox_4_6, 'Value') == 0)
    set(handles.type_s_4_6, 'String', {'(Closed)'});
end


% --- Executes on button press in checkbox_4_7.
function checkbox_4_7_Callback(hObject, eventdata, handles)
% hObject    handle to checkbox_4_7 (see GCBO)
% eventdata  reserved - to be defined in a future version of MATLAB
% handles    structure with handles and user data (see GUIDATA)

% Hint: get(hObject,'Value') returns toggle state of checkbox_4_7
load dtw_deice_data.mat;
if (get(handles.checkbox_4_7, 'Value') == 1)
    set(handles.type_s_4_7, 'String', ac_list_s_4_7);
elseif (get(handles.checkbox_4_7, 'Value') == 0)
    set(handles.type_s_4_7, 'String', {'(Closed)'});
end


% --- Executes on button press in checkbox_3_1.
function checkbox_3_1_Callback(hObject, eventdata, handles)
% hObject    handle to checkbox_3_1 (see GCBO)
% eventdata  reserved - to be defined in a future version of MATLAB
% handles    structure with handles and user data (see GUIDATA)
```

```matlab
% Hint: get(hObject,'Value') returns toggle state of checkbox_3_1
load dtw_deice_data.mat;
if (get(handles.checkbox_3_1, 'Value') == 1)
set(handles.type_s_3_1, 'String', ac_list_s_3_1);
elseif (get(handles.checkbox_3_1, 'Value') == 0)
set(handles.type_s_3_1, 'String', {'(Closed)'});
end


% --- Executes on button press in checkbox_3_2.
function checkbox_3_2_Callback(hObject, eventdata, handles)
% hObject handle to checkbox_3_2 (see GCBO)
% eventdata reserved - to be defined in a future version of MATLAB
% handles structure with handles and user data (see GUIDATA)

% Hint: get(hObject,'Value') returns toggle state of checkbox_3_2
load dtw_deice_data.mat;
if (get(handles.checkbox_3_2, 'Value') == 1)
set(handles.type_s_3_2, 'String', ac_list_s_3_2);
elseif (get(handles.checkbox_3_2, 'Value') == 0)
set(handles.type_s_3_2, 'String', {'(Closed)'});
end


% --- Executes on button press in checkbox_3_3.
function checkbox_3_3_Callback(hObject, eventdata, handles)
% hObject handle to checkbox_3_3 (see GCBO)
% eventdata reserved - to be defined in a future version of MATLAB
% handles structure with handles and user data (see GUIDATA)

% Hint: get(hObject,'Value') returns toggle state of checkbox_3_3
load dtw_deice_data.mat;
if (get(handles.checkbox_3_3, 'Value') == 1)
set(handles.type_s_3_3, 'String', ac_list_s_3_3);
elseif (get(handles.checkbox_3_3, 'Value') == 0)
set(handles.type_s_3_3, 'String', {'(Closed)'});
end


% --- Executes on button press in checkbox_3_4.
function checkbox_3_4_Callback(hObject, eventdata, handles)
% hObject handle to checkbox_3_4 (see GCBO)
% eventdata reserved - to be defined in a future version of MATLAB
% handles structure with handles and user data (see GUIDATA)

% Hint: get(hObject,'Value') returns toggle state of checkbox_3_4
load dtw_deice_data.mat;
if (get(handles.checkbox_3_4, 'Value') == 1)
set(handles.type_s_3_4, 'String', ac_list_s_3_4);
elseif (get(handles.checkbox_3_4, 'Value') == 0)
set(handles.type_s_3_4, 'String', {'(Closed)'});
end


% --- Executes on button press in checkbox_3_5.
function checkbox_3_5_Callback(hObject, eventdata, handles)
% hObject handle to checkbox_3_5 (see GCBO)
% eventdata reserved - to be defined in a future version of MATLAB
% handles structure with handles and user data (see GUIDATA)

% Hint: get(hObject,'Value') returns toggle state of checkbox_3_5
load dtw_deice_data.mat;
if (get(handles.checkbox_3_5, 'Value') == 1)
set(handles.type_s_3_5, 'String', ac_list_s_3_5);
```

```
552 elseif (get(handles.checkbox_3_5, 'Value') == 0)
553 set(handles.type_s_3_5, 'String', {'(Closed)'});
554 end
555
556
557 % --- Executes on button press in checkbox_3_6.
558 function checkbox_3_6_Callback(hObject, eventdata, handles)
559 % hObject handle to checkbox_3_6 (see GCBO)
560 % eventdata reserved - to be defined in a future version of MATLAB
561 % handles structure with handles and user data (see GUIDATA)
562
563 % Hint: get(hObject,'Value') returns toggle state of checkbox_3_6
564 load dtw_deice_data.mat;
565 if (get(handles.checkbox_3_6, 'Value') == 1)
566 set(handles.type_s_3_6, 'String', ac_list_s_3_6);
567 elseif (get(handles.checkbox_3_6, 'Value') == 0)
568 set(handles.type_s_3_6, 'String', {'(Closed)'});
569 end
570
571
572 % --- Executes on button press in checkbox_3_7.
573 function checkbox_3_7_Callback(hObject, eventdata, handles)
574 % hObject handle to checkbox_3_7 (see GCBO)
575 % eventdata reserved - to be defined in a future version of MATLAB
576 % handles structure with handles and user data (see GUIDATA)
577
578 % Hint: get(hObject,'Value') returns toggle state of checkbox_3_7
579 load dtw_deice_data.mat;
580 if (get(handles.checkbox_3_7, 'Value') == 1)
581 set(handles.type_s_3_7, 'String', ac_list_s_3_7);
582 elseif (get(handles.checkbox_3_7, 'Value') == 0)
583 set(handles.type_s_3_7, 'String', {'(Closed)'});
584 end
585
586
587
588 % --- Executes during object creation, after setting all properties.
589 function type_s_4_1_CreateFcn(hObject, eventdata, handles)
590 % hObject handle to type_s_4_1 (see GCBO)
591 % eventdata reserved - to be defined in a future version of MATLAB
592 % handles empty - handles not created until after all CreateFcns called
593
594 % Hint: popupmenu controls usually have a white background on Windows.
595 %        See ISPC and COMPUTER.
596 if ispc
597 set(hObject,'BackgroundColor','white');
598 else
599 set(hObject,'BackgroundColor',get(0,'defaultUicontrolBackgroundColor'));
600 end
601
602 % --- Executes on selection change in type_s_4_1.
603 function type_s_4_1_Callback(hObject, eventdata, handles)
604 % hObject handle to type_s_4_1 (see GCBO)
605 % eventdata reserved - to be defined in a future version of MATLAB
606 % handles structure with handles and user data (see GUIDATA)
607
608 % Hints: contents = get(hObject,'String') returns type_s_4_1 contents as cell array
609 %        contents{get(hObject,'Value')} returns selected item from type_s_4_1
610
611
612
613 % --- Executes during object creation, after setting all properties.
```

```
614 function type_s_4_2_CreateFcn(hObject, eventdata, handles)
615 % hObject handle to type_s_4_2 (see GCBO)
616 % eventdata reserved - to be defined in a future version of MATLAB
617 % handles empty - handles not created until after all CreateFcns called
618
619 % Hint: popupmenu controls usually have a white background on Windows.
620 %       See ISPC and COMPUTER.
621 if ispc
622 set(hObject,'BackgroundColor','white');
623 else
624 set(hObject,'BackgroundColor',get(0,'defaultUicontrolBackgroundColor'));
625 end
626
627 % --- Executes on selection change in type_s_4_2.
628 function type_s_4_2_Callback(hObject, eventdata, handles)
629 % hObject handle to type_s_4_2 (see GCBO)
630 % eventdata reserved - to be defined in a future version of MATLAB
631 % handles structure with handles and user data (see GUIDATA)
632
633 % Hints: contents = get(hObject,'String') returns type_s_4_2 contents as cell array
634 %        contents{get(hObject,'Value')} returns selected item from type_s_4_2
635
636
637
638 % --- Executes during object creation, after setting all properties.
639 function type_s_4_3_CreateFcn(hObject, eventdata, handles)
640 % hObject handle to type_s_4_3 (see GCBO)
641 % eventdata reserved - to be defined in a future version of MATLAB
642 % handles empty - handles not created until after all CreateFcns called
643
644 % Hint: popupmenu controls usually have a white background on Windows.
645 %       See ISPC and COMPUTER.
646 if ispc
647 set(hObject,'BackgroundColor','white');
648 else
649 set(hObject,'BackgroundColor',get(0,'defaultUicontrolBackgroundColor'));
650 end
651
652 % --- Executes on selection change in type_s_4_3.
653 function type_s_4_3_Callback(hObject, eventdata, handles)
654 % hObject handle to type_s_4_3 (see GCBO)
655 % eventdata reserved - to be defined in a future version of MATLAB
656 % handles structure with handles and user data (see GUIDATA)
657
658 % Hints: contents = get(hObject,'String') returns type_s_4_3 contents as cell array
659 %        contents{get(hObject,'Value')} returns selected item from type_s_4_3
660
661
662
663 % --- Executes during object creation, after setting all properties.
664 function type_s_4_4_CreateFcn(hObject, eventdata, handles)
665 % hObject handle to type_s_4_4 (see GCBO)
666 % eventdata reserved - to be defined in a future version of MATLAB
667 % handles empty - handles not created until after all CreateFcns called
668
669 % Hint: popupmenu controls usually have a white background on Windows.
670 %       See ISPC and COMPUTER.
671 if ispc
672 set(hObject,'BackgroundColor','white');
673 else
```

```
674 set(hObject,'BackgroundColor',get(0,'defaultUicontrolBackgroundColor'));
675 end
676
677 % --- Executes on selection change in type_s_4_4.
678 function type_s_4_4_Callback(hObject, eventdata, handles)
679 % hObject handle to type_s_4_4 (see GCBO)
680 % eventdata reserved - to be defined in a future version of MATLAB
681 % handles structure with handles and user data (see GUIDATA)
682
683 % Hints: contents = get(hObject,'String') returns type_s_4_4 contents as cell array
684 % contents{get(hObject,'Value')} returns selected item from type_s_4_4
685
686
687
688 % --- Executes during object creation, after setting all properties.
689 function type_s_4_5_CreateFcn(hObject, eventdata, handles)
690 % hObject handle to type_s_4_5 (see GCBO)
691 % eventdata reserved - to be defined in a future version of MATLAB
692 % handles empty - handles not created until after all CreateFcns called
693
694 % Hint: popupmenu controls usually have a white background on Windows.
695 %       See ISPC and COMPUTER.
696 if ispc
697 set(hObject,'BackgroundColor','white');
698 else
699 set(hObject,'BackgroundColor',get(0,'defaultUicontrolBackgroundColor'));
700 end
701
702 % --- Executes on selection change in type_s_4_5.
703 function type_s_4_5_Callback(hObject, eventdata, handles)
704 % hObject handle to type_s_4_5 (see GCBO)
705 % eventdata reserved - to be defined in a future version of MATLAB
706 % handles structure with handles and user data (see GUIDATA)
707
708 % Hints: contents = get(hObject,'String') returns type_s_4_5 contents as cell array
709 % contents{get(hObject,'Value')} returns selected item from type_s_4_5
710
711
712
713 % --- Executes during object creation, after setting all properties.
714 function type_s_4_6_CreateFcn(hObject, eventdata, handles)
715 % hObject handle to type_s_4_6 (see GCBO)
716 % eventdata reserved - to be defined in a future version of MATLAB
717 % handles empty - handles not created until after all CreateFcns called
718
719 % Hint: popupmenu controls usually have a white background on Windows.
720 %       See ISPC and COMPUTER.
721 if ispc
722 set(hObject,'BackgroundColor','white');
723 else
724 set(hObject,'BackgroundColor',get(0,'defaultUicontrolBackgroundColor'));
725 end
726
727 % --- Executes on selection change in type_s_4_6.
728 function type_s_4_6_Callback(hObject, eventdata, handles)
729 % hObject handle to type_s_4_6 (see GCBO)
730 % eventdata reserved - to be defined in a future version of MATLAB
731 % handles structure with handles and user data (see GUIDATA)
732
```

```matlab
% Hints: contents = get(hObject,'String') returns type_s_4_6 contents as cell array
%        contents{get(hObject,'Value')} returns selected item from type_s_4_6


% --- Executes during object creation, after setting all properties.
function type_s_4_7_CreateFcn(hObject, eventdata, handles)
% hObject    handle to type_s_4_7 (see GCBO)
% eventdata  reserved - to be defined in a future version of MATLAB
% handles    empty - handles not created until after all CreateFcns called

% Hint: popupmenu controls usually have a white background on Windows.
%       See ISPC and COMPUTER.
if ispc
    set(hObject,'BackgroundColor','white');
else
    set(hObject,'BackgroundColor',get(0,'defaultUicontrolBackgroundColor'));
end

% --- Executes on selection change in type_s_4_7.
function type_s_4_7_Callback(hObject, eventdata, handles)
% hObject    handle to type_s_4_7 (see GCBO)
% eventdata  reserved - to be defined in a future version of MATLAB
% handles    structure with handles and user data (see GUIDATA)

% Hints: contents = get(hObject,'String') returns type_s_4_7 contents as cell array
%        contents{get(hObject,'Value')} returns selected item from type_s_4_7



% --- Executes during object creation, after setting all properties.
function type_s_3_1_CreateFcn(hObject, eventdata, handles)
% hObject    handle to type_s_3_1 (see GCBO)
% eventdata  reserved - to be defined in a future version of MATLAB
% handles    empty - handles not created until after all CreateFcns called

% Hint: popupmenu controls usually have a white background on Windows.
%       See ISPC and COMPUTER.
if ispc
    set(hObject,'BackgroundColor','white');
else
    set(hObject,'BackgroundColor',get(0,'defaultUicontrolBackgroundColor'));
end

% --- Executes on selection change in type_s_3_1.
function type_s_3_1_Callback(hObject, eventdata, handles)
% hObject    handle to type_s_3_1 (see GCBO)
% eventdata  reserved - to be defined in a future version of MATLAB
% handles    structure with handles and user data (see GUIDATA)

% Hints: contents = get(hObject,'String') returns type_s_3_1 contents as cell array
%        contents{get(hObject,'Value')} returns selected item from type_s_3_1



% --- Executes during object creation, after setting all properties.
function type_s_3_2_CreateFcn(hObject, eventdata, handles)
% hObject    handle to type_s_3_2 (see GCBO)
% eventdata  reserved - to be defined in a future version of MATLAB
```

```
791 % handles empty - handles not created until after all CreateFcns called
792
793 % Hint: popupmenu controls usually have a white background on Windows.
794 % See ISPC and COMPUTER.
795 if ispc
796 set(hObject,'BackgroundColor','white');
797 else
798 set(hObject,'BackgroundColor',get(0,'defaultUicontrolBackgroundColor'));
799 end
800
801 % --- Executes on selection change in type_s_3_2.
802 function type_s_3_2_Callback(hObject, eventdata, handles)
803 % hObject handle to type_s_3_2 (see GCBO)
804 % eventdata reserved - to be defined in a future version of MATLAB
805 % handles structure with handles and user data (see GUIDATA)
806
807 % Hints: contents = get(hObject,'String') returns type_s_3_2 contents as cell array
808 % contents{get(hObject,'Value')} returns selected item from type_s_3_2
809
810
811
812 % --- Executes during object creation, after setting all properties.
813 function type_s_3_3_CreateFcn(hObject, eventdata, handles)
814 % hObject handle to type_s_3_3 (see GCBO)
815 % eventdata reserved - to be defined in a future version of MATLAB
816 % handles empty - handles not created until after all CreateFcns called
817
818 % Hint: popupmenu controls usually have a white background on Windows.
819 % See ISPC and COMPUTER.
820 if ispc
821 set(hObject,'BackgroundColor','white');
822 else
823 set(hObject,'BackgroundColor',get(0,'defaultUicontrolBackgroundColor'));
824 end
825
826 % --- Executes on selection change in type_s_3_3.
827 function type_s_3_3_Callback(hObject, eventdata, handles)
828 % hObject handle to type_s_3_3 (see GCBO)
829 % eventdata reserved - to be defined in a future version of MATLAB
830 % handles structure with handles and user data (see GUIDATA)
831
832 % Hints: contents = get(hObject,'String') returns type_s_3_3 contents as cell array
833 % contents{get(hObject,'Value')} returns selected item from type_s_3_3
834
835
836
837 % --- Executes during object creation, after setting all properties.
838 function type_s_3_4_CreateFcn(hObject, eventdata, handles)
839 % hObject handle to type_s_3_4 (see GCBO)
840 % eventdata reserved - to be defined in a future version of MATLAB
841 % handles empty - handles not created until after all CreateFcns called
842
843 % Hint: popupmenu controls usually have a white background on Windows.
844 % See ISPC and COMPUTER.
845 if ispc
846 set(hObject,'BackgroundColor','white');
847 else
848 set(hObject,'BackgroundColor',get(0,'defaultUicontrolBackgroundColor'));
849 end
850
```

```matlab
% --- Executes on selection change in type_s_3_4.
function type_s_3_4_Callback(hObject, eventdata, handles)
% hObject  handle to type_s_3_4 (see GCBO)
% eventdata  reserved - to be defined in a future version of MATLAB
% handles  structure with handles and user data (see GUIDATA)

% Hints: contents = get(hObject,'String') returns type_s_3_4 contents as cell array
%        contents{get(hObject,'Value')} returns selected item from type_s_3_4



% --- Executes during object creation, after setting all properties.
function type_s_3_5_CreateFcn(hObject, eventdata, handles)
% hObject  handle to type_s_3_5 (see GCBO)
% eventdata  reserved - to be defined in a future version of MATLAB
% handles  empty - handles not created until after all CreateFcns called

% Hint: popupmenu controls usually have a white background on Windows.
%       See ISPC and COMPUTER.
if ispc
    set(hObject,'BackgroundColor','white');
else
    set(hObject,'BackgroundColor',get(0,'defaultUicontrolBackgroundColor'));
end

% --- Executes on selection change in type_s_3_5.
function type_s_3_5_Callback(hObject, eventdata, handles)
% hObject  handle to type_s_3_5 (see GCBO)
% eventdata  reserved - to be defined in a future version of MATLAB
% handles  structure with handles and user data (see GUIDATA)

% Hints: contents = get(hObject,'String') returns type_s_3_5 contents as cell array
%        contents{get(hObject,'Value')} returns selected item from type_s_3_5




% --- Executes during object creation, after setting all properties.
function type_s_3_6_CreateFcn(hObject, eventdata, handles)
% hObject  handle to type_s_3_6 (see GCBO)
% eventdata  reserved - to be defined in a future version of MATLAB
% handles  empty - handles not created until after all CreateFcns called

% Hint: popupmenu controls usually have a white background on Windows.
%       See ISPC and COMPUTER.
if ispc
    set(hObject,'BackgroundColor','white');
else
    set(hObject,'BackgroundColor',get(0,'defaultUicontrolBackgroundColor'));
end

% --- Executes on selection change in type_s_3_6.
function type_s_3_6_Callback(hObject, eventdata, handles)
% hObject  handle to type_s_3_6 (see GCBO)
% eventdata  reserved - to be defined in a future version of MATLAB
% handles  structure with handles and user data (see GUIDATA)

% Hints: contents = get(hObject,'String') returns type_s_3_6 contents as cell array
%        contents{get(hObject,'Value')} returns selected item from type_s_3_6
```

```matlab
% --- Executes during object creation, after setting all properties.
function type_s_3_7_CreateFcn(hObject, eventdata, handles)
% hObject    handle to type_s_3_7 (see GCBO)
% eventdata  reserved - to be defined in a future version of MATLAB
% handles    empty - handles not created until after all CreateFcns called

% Hint: popupmenu controls usually have a white background on Windows.
%       See ISPC and COMPUTER.
if ispc
    set(hObject,'BackgroundColor','white');
else
    set(hObject,'BackgroundColor',get(0,'defaultUicontrolBackgroundColor'));
end


% --- Executes on selection change in type_s_3_7.
function type_s_3_7_Callback(hObject, eventdata, handles)
% hObject    handle to type_s_3_7 (see GCBO)
% eventdata  reserved - to be defined in a future version of MATLAB
% handles    structure with handles and user data (see GUIDATA)

% Hints: contents = get(hObject,'String') returns type_s_3_7 contents as cell array
%        contents{get(hObject,'Value')} returns selected item from type_s_3_7



% --- Executes during object creation, after setting all properties.
function time_s_4_1_CreateFcn(hObject, eventdata, handles)
% hObject    handle to time_dc9_b757_s_4_1 (see GCBO)
% eventdata  reserved - to be defined in a future version of MATLAB
% handles    empty - handles not created until after all CreateFcns called

% Hint: edit controls usually have a white background on Windows.
%       See ISPC and COMPUTER.
if ispc
    set(hObject,'BackgroundColor','white');
else
    set(hObject,'BackgroundColor',get(0,'defaultUicontrolBackgroundColor'));
end

function time_s_4_1_Callback(hObject, eventdata, handles)
% hObject    handle to time_dc9_b757_s_4_1 (see GCBO)
% eventdata  reserved - to be defined in a future version of MATLAB
% handles    structure with handles and user data (see GUIDATA)

% Hints: get(hObject,'String') returns contents of time_dc9_b757_s_4_1 as text
%        str2double(get(hObject,'String')) returns contents of time_dc9_b757_s_4_1 as a double



% --- Executes during object creation, after setting all properties.
function time_s_4_2_CreateFcn(hObject, eventdata, handles)
% hObject    handle to time_dc9_b757_s_4_2 (see GCBO)
% eventdata  reserved - to be defined in a future version of MATLAB
% handles    empty - handles not created until after all CreateFcns called

% Hint: edit controls usually have a white background on Windows.
%       See ISPC and COMPUTER.
if ispc
```

```
969 set(hObject,'BackgroundColor','white');
970 else
971 set(hObject,'BackgroundColor',get(0,'defaultUicontrolBackgroundColor'));
972 end
973
974 function time_s_4_2_Callback(hObject, eventdata, handles)
975 % hObject handle to time_dc9_b757_s_4_2 (see GCBO)
976 % eventdata reserved - to be defined in a future version of MATLAB
977 % handles structure with handles and user data (see GUIDATA)
978
979 % Hints: get(hObject,'String') returns contents of time_dc9_b757_s_4_2 as text
980 % str2double(get(hObject,'String')) returns contents of time_dc9_b757_s_4_2 as a double
981
982
983
984 % --- Executes during object creation, after setting all properties.
985 function time_s_4_3_CreateFcn(hObject, eventdata, handles)
986 % hObject handle to time_dc9_b757_s_4_3 (see GCBO)
987 % eventdata reserved - to be defined in a future version of MATLAB
988 % handles empty - handles not created until after all CreateFcns called
989
990 % Hint: edit controls usually have a white background on Windows.
991 % See ISPC and COMPUTER.
992 if ispc
993 set(hObject,'BackgroundColor','white');
994 else
995 set(hObject,'BackgroundColor',get(0,'defaultUicontrolBackgroundColor'));
996 end
997
998 function time_s_4_3_Callback(hObject, eventdata, handles)
999 % hObject handle to time_dc9_b757_s_4_3 (see GCBO)
1000 % eventdata reserved - to be defined in a future version of MATLAB
1001 % handles structure with handles and user data (see GUIDATA)
1002
1003 % Hints: get(hObject,'String') returns contents of time_dc9_b757_s_4_3 as text
1004 % str2double(get(hObject,'String')) returns contents of time_dc9_b757_s_4_3 as a double
1005
1006
1007
1008 % --- Executes during object creation, after setting all properties.
1009 function time_s_4_4_CreateFcn(hObject, eventdata, handles)
1010 % hObject handle to time_dc9_b757_s_4_4 (see GCBO)
1011 % eventdata reserved - to be defined in a future version of MATLAB
1012 % handles empty - handles not created until after all CreateFcns called
1013
1014 % Hint: edit controls usually have a white background on Windows.
1015 % See ISPC and COMPUTER.
1016 if ispc
1017 set(hObject,'BackgroundColor','white');
1018 else
1019 set(hObject,'BackgroundColor',get(0,'defaultUicontrolBackgroundColor'));
1020 end
1021
1022 function time_s_4_4_Callback(hObject, eventdata, handles)
1023 % hObject handle to time_dc9_b757_s_4_4 (see GCBO)
1024 % eventdata reserved - to be defined in a future version of MATLAB
1025 % handles structure with handles and user data (see GUIDATA)
1026
1027 % Hints: get(hObject,'String') returns contents of time_dc9_b757_s_4_4 as text
```

```matlab
1028 % str2double(get(hObject,'String')) returns contents of time_dc9_b757_s_4_4 as a double
1029
1030
1031
1032 % --- Executes during object creation, after setting all properties.
1033 function time_s_4_5_CreateFcn(hObject, eventdata, handles)
1034 % hObject    handle to time_dc9_b757_s_4_5 (see GCBO)
1035 % eventdata  reserved - to be defined in a future version of MATLAB
1036 % handles    empty - handles not created until after all CreateFcns called
1037
1038 % Hint: edit controls usually have a white background on Windows.
1039 %       See ISPC and COMPUTER.
1040 if ispc
1041     set(hObject,'BackgroundColor','white');
1042 else
1043     set(hObject,'BackgroundColor',get(0,'defaultUicontrolBackgroundColor'));
1044 end
1045
1046 function time_s_4_5_Callback(hObject, eventdata, handles)
1047 % hObject    handle to time_dc9_b757_s_4_5 (see GCBO)
1048 % eventdata  reserved - to be defined in a future version of MATLAB
1049 % handles    structure with handles and user data (see GUIDATA)
1050
1051 % Hints: get(hObject,'String') returns contents of time_dc9_b757_s_4_5 as text
1052 % str2double(get(hObject,'String')) returns contents of time_dc9_b757_s_4_5 as a double
1053
1054
1055
1056 % --- Executes during object creation, after setting all properties.
1057 function time_s_4_6_CreateFcn(hObject, eventdata, handles)
1058 % hObject    handle to time_dc9_b757_s_4_6 (see GCBO)
1059 % eventdata  reserved - to be defined in a future version of MATLAB
1060 % handles    empty - handles not created until after all CreateFcns called
1061
1062 % Hint: edit controls usually have a white background on Windows.
1063 %       See ISPC and COMPUTER.
1064 if ispc
1065     set(hObject,'BackgroundColor','white');
1066 else
1067     set(hObject,'BackgroundColor',get(0,'defaultUicontrolBackgroundColor'));
1068 end
1069
1070 function time_s_4_6_Callback(hObject, eventdata, handles)
1071 % hObject    handle to time_dc9_b757_s_4_6 (see GCBO)
1072 % eventdata  reserved - to be defined in a future version of MATLAB
1073 % handles    structure with handles and user data (see GUIDATA)
1074
1075 % Hints: get(hObject,'String') returns contents of time_dc9_b757_s_4_6 as text
1076 % str2double(get(hObject,'String')) returns contents of time_dc9_b757_s_4_6 as a double
1077
1078
1079 % --- Executes during object creation, after setting all properties.
1080 function time_s_4_7_CreateFcn(hObject, eventdata, handles)
1081 % hObject    handle to time_dc9_b757_s_4_7 (see GCBO)
1082 % eventdata  reserved - to be defined in a future version of MATLAB
1083 % handles    empty - handles not created until after all CreateFcns called
1084
1085 % Hint: edit controls usually have a white background on Windows.
1086 %       See ISPC and COMPUTER.
```

```matlab
1087 if ispc
1088 set(hObject,'BackgroundColor','white');
1089 else
1090 set(hObject,'BackgroundColor',get(0,'defaultUicontrolBackgroundColor'));
1091 end
1092
1093 function time_s_4_7_Callback(hObject, eventdata, handles)
1094 % hObject handle to time_dc9_b757_s_4_7 (see GCBO)
1095 % eventdata reserved - to be defined in a future version of MATLAB
1096 % handles structure with handles and user data (see GUIDATA)
1097
1098 % Hints: get(hObject,'String') returns contents of time_dc9_b757_s_4_7 as text
1099 % str2double(get(hObject,'String')) returns contents of time_dc9_b757_s_4_7 as a double
1100
1101
1102
1103 % --- Executes during object creation, after setting all properties.
1104 function time_s_3_1_CreateFcn(hObject, eventdata, handles)
1105 % hObject handle to time_dc9_b757_s_3_1 (see GCBO)
1106 % eventdata reserved - to be defined in a future version of MATLAB
1107 % handles empty - handles not created until after all CreateFcns called
1108
1109 % Hint: edit controls usually have a white background on Windows.
1110 % See ISPC and COMPUTER.
1111 if ispc
1112 set(hObject,'BackgroundColor','white');
1113 else
1114 set(hObject,'BackgroundColor',get(0,'defaultUicontrolBackgroundColor'));
1115 end
1116
1117 function time_s_3_1_Callback(hObject, eventdata, handles)
1118 % hObject handle to time_dc9_b757_s_3_1 (see GCBO)
1119 % eventdata reserved - to be defined in a future version of MATLAB
1120 % handles structure with handles and user data (see GUIDATA)
1121
1122 % Hints: get(hObject,'String') returns contents of time_dc9_b757_s_3_1 as text
1123 % str2double(get(hObject,'String')) returns contents of time_dc9_b757_s_3_1 as a double
1124
1125
1126
1127 % --- Executes during object creation, after setting all properties.
1128 function time_s_3_2_CreateFcn(hObject, eventdata, handles)
1129 % hObject handle to time_s_3_2 (see GCBO)
1130 % eventdata reserved - to be defined in a future version of MATLAB
1131 % handles empty - handles not created until after all CreateFcns called
1132
1133 % Hint: edit controls usually have a white background on Windows.
1134 % See ISPC and COMPUTER.
1135 if ispc
1136 set(hObject,'BackgroundColor','white');
1137 else
1138 set(hObject,'BackgroundColor',get(0,'defaultUicontrolBackgroundColor'));
1139 end
1140
1141 function time_s_3_2_Callback(hObject, eventdata, handles)
1142 % hObject handle to time_s_3_2 (see GCBO)
1143 % eventdata reserved - to be defined in a future version of MATLAB
1144 % handles structure with handles and user data (see GUIDATA)
1145
1146 % Hints: get(hObject,'String') returns contents of time_s_3_2 as text
```

```
1147 % str2double(get(hObject,'String')) returns contents of time_s_3_2 as a double
1148
1149
1150
1151 % --- Executes during object creation, after setting all properties.
1152 function time_s_3_3_CreateFcn(hObject, eventdata, handles)
1153 % hObject    handle to time_s_3_3 (see GCBO)
1154 % eventdata  reserved - to be defined in a future version of MATLAB
1155 % handles    empty - handles not created until after all CreateFcns called
1156
1157 % Hint: edit controls usually have a white background on Windows.
1158 %       See ISPC and COMPUTER.
1159 if ispc
1160     set(hObject,'BackgroundColor','white');
1161 else
1162     set(hObject,'BackgroundColor',get(0,'defaultUicontrolBackgroundColor'));
1163 end
1164
1165 function time_s_3_3_Callback(hObject, eventdata, handles)
1166 % hObject    handle to time_s_3_3 (see GCBO)
1167 % eventdata  reserved - to be defined in a future version of MATLAB
1168 % handles    structure with handles and user data (see GUIDATA)
1169
1170 % Hints: get(hObject,'String') returns contents of time_s_3_3 as text
1171 %        str2double(get(hObject,'String')) returns contents of time_s_3_3 as a double
1172
1173
1174
1175 % --- Executes during object creation, after setting all properties.
1176 function time_s_3_4_CreateFcn(hObject, eventdata, handles)
1177 % hObject    handle to time_s_3_4 (see GCBO)
1178 % eventdata  reserved - to be defined in a future version of MATLAB
1179 % handles    empty - handles not created until after all CreateFcns called
1180
1181 % Hint: edit controls usually have a white background on Windows.
1182 %       See ISPC and COMPUTER.
1183 if ispc
1184     set(hObject,'BackgroundColor','white');
1185 else
1186     set(hObject,'BackgroundColor',get(0,'defaultUicontrolBackgroundColor'));
1187 end
1188
1189 function time_s_3_4_Callback(hObject, eventdata, handles)
1190 % hObject    handle to time_s_3_4 (see GCBO)
1191 % eventdata  reserved - to be defined in a future version of MATLAB
1192 % handles    structure with handles and user data (see GUIDATA)
1193
1194 % Hints: get(hObject,'String') returns contents of time_s_3_4 as text
1195 %        str2double(get(hObject,'String')) returns contents of time_s_3_4 as a double
1196
1197
1198
1199 % --- Executes during object creation, after setting all properties.
1200 function time_s_3_5_CreateFcn(hObject, eventdata, handles)
1201 % hObject    handle to time_s_3_5 (see GCBO)
1202 % eventdata  reserved - to be defined in a future version of MATLAB
1203 % handles    empty - handles not created until after all CreateFcns called
1204
1205 % Hint: edit controls usually have a white background on Windows.
1206 %       See ISPC and COMPUTER.
1207 if ispc
```

```
1208 set(hObject,'BackgroundColor','white');
1209 else
1210 set(hObject,'BackgroundColor',get(0,'defaultUicontrolBackgroundColor'));
1211 end
1212
1213 function time_s_3_5_Callback(hObject, eventdata, handles)
1214 % hObject handle to time_s_3_5 (see GCBO)
1215 % eventdata reserved - to be defined in a future version of MATLAB
1216 % handles structure with handles and user data (see GUIDATA)
1217
1218 % Hints: get(hObject,'String') returns contents of time_s_3_5 as text
1219 %        str2double(get(hObject,'String')) returns contents of time_s_3_5 as a double
1220
1221
1222
1223 % --- Executes during object creation, after setting all properties.
1224 function time_s_3_6_CreateFcn(hObject, eventdata, handles)
1225 % hObject handle to time_s_3_6 (see GCBO)
1226 % eventdata reserved - to be defined in a future version of MATLAB
1227 % handles empty - handles not created until after all CreateFcns called
1228
1229 % Hint: edit controls usually have a white background on Windows.
1230 %       See ISPC and COMPUTER.
1231 if ispc
1232 set(hObject,'BackgroundColor','white');
1233 else
1234 set(hObject,'BackgroundColor',get(0,'defaultUicontrolBackgroundColor'));
1235 end
1236
1237 function time_s_3_6_Callback(hObject, eventdata, handles)
1238 % hObject handle to time_s_3_6 (see GCBO)
1239 % eventdata reserved - to be defined in a future version of MATLAB
1240 % handles structure with handles and user data (see GUIDATA)
1241
1242 % Hints: get(hObject,'String') returns contents of time_s_3_6 as text
1243 %        str2double(get(hObject,'String')) returns contents of time_s_3_6 as a double
1244
1245
1246 % --- Executes during object creation, after setting all properties.
1247 function time_s_3_7_CreateFcn(hObject, eventdata, handles)
1248 % hObject handle to time_s_3_7 (see GCBO)
1249 % eventdata reserved - to be defined in a future version of MATLAB
1250 % handles empty - handles not created until after all CreateFcns called
1251
1252 % Hint: edit controls usually have a white background on Windows.
1253 %       See ISPC and COMPUTER.
1254 if ispc
1255 set(hObject,'BackgroundColor','white');
1256 else
1257 set(hObject,'BackgroundColor',get(0,'defaultUicontrolBackgroundColor'));
1258 end
1259
1260 function time_s_3_7_Callback(hObject, eventdata, handles)
1261 % hObject handle to time_s_3_7 (see GCBO)
1262 % eventdata reserved - to be defined in a future version of MATLAB
1263 % handles structure with handles and user data (see GUIDATA)
1264
1265 % Hints: get(hObject,'String') returns contents of time_s_3_7 as text
1266 %        str2double(get(hObject,'String')) returns contents of time_s_3_7 as a double
1267
1268
```

```matlab
% --- Executes during object creation, after setting all properties.
function type_q_4_1_CreateFcn(hObject, eventdata, handles)
% hObject    handle to type_q_4_1 (see GCBO)
% eventdata  reserved - to be defined in a future version of MATLAB
% handles    empty - handles not created until after all CreateFcns called

% Hint: popupmenu controls usually have a white background on Windows.
%       See ISPC and COMPUTER.
if ispc
    set(hObject,'BackgroundColor','white');
else
    set(hObject,'BackgroundColor',get(0,'defaultUicontrolBackgroundColor'));
end

% --- Executes on selection change in type_q_4_1.
function type_q_4_1_Callback(hObject, eventdata, handles)
% hObject    handle to type_q_4_1 (see GCBO)
% eventdata  reserved - to be defined in a future version of MATLAB
% handles    structure with handles and user data (see GUIDATA)

% Hints: contents = get(hObject,'String') returns type_q_4_1 contents as cell array
%        contents{get(hObject,'Value')} returns selected item from type_q_4_1


% --- Executes during object creation, after setting all properties.
function type_q_4_2_CreateFcn(hObject, eventdata, handles)
% hObject    handle to type_q_4_2 (see GCBO)
% eventdata  reserved - to be defined in a future version of MATLAB
% handles    empty - handles not created until after all CreateFcns called

% Hint: popupmenu controls usually have a white background on Windows.
%       See ISPC and COMPUTER.
if ispc
    set(hObject,'BackgroundColor','white');
else
    set(hObject,'BackgroundColor',get(0,'defaultUicontrolBackgroundColor'));
end

% --- Executes on selection change in type_q_4_2.
function type_q_4_2_Callback(hObject, eventdata, handles)
% hObject    handle to type_q_4_2 (see GCBO)
% eventdata  reserved - to be defined in a future version of MATLAB
% handles    structure with handles and user data (see GUIDATA)

% Hints: contents = get(hObject,'String') returns type_q_4_2 contents as cell array
%        contents{get(hObject,'Value')} returns selected item from type_q_4_2


% --- Executes during object creation, after setting all properties.
function type_q_4_3_CreateFcn(hObject, eventdata, handles)
% hObject    handle to type_q_4_3 (see GCBO)
% eventdata  reserved - to be defined in a future version of MATLAB
% handles    empty - handles not created until after all CreateFcns called

% Hint: popupmenu controls usually have a white background on Windows.
%       See ISPC and COMPUTER.
if ispc
    set(hObject,'BackgroundColor','white');
else
```

```matlab
set(hObject,'BackgroundColor',get(0,'defaultUicontrolBackgroundColor'));
end

% --- Executes on selection change in type_q_4_3.
function type_q_4_3_Callback(hObject, eventdata, handles)
% hObject    handle to type_q_4_3 (see GCBO)
% eventdata  reserved - to be defined in a future version of MATLAB
% handles    structure with handles and user data (see GUIDATA)

% Hints: contents = get(hObject,'String') returns type_q_4_3 contents as cell array
%        contents{get(hObject,'Value')} returns selected item from type_q_4_3


% --- Executes during object creation, after setting all properties.
function type_q_4_4_CreateFcn(hObject, eventdata, handles)
% hObject    handle to type_q_4_4 (see GCBO)
% eventdata  reserved - to be defined in a future version of MATLAB
% handles    empty - handles not created until after all CreateFcns called

% Hint: popupmenu controls usually have a white background on Windows.
%       See ISPC and COMPUTER.
if ispc
set(hObject,'BackgroundColor','white');
else
set(hObject,'BackgroundColor',get(0,'defaultUicontrolBackgroundColor'));
end

% --- Executes on selection change in type_q_4_4.
function type_q_4_4_Callback(hObject, eventdata, handles)
% hObject    handle to type_q_4_4 (see GCBO)
% eventdata  reserved - to be defined in a future version of MATLAB
% handles    structure with handles and user data (see GUIDATA)

% Hints: contents = get(hObject,'String') returns type_q_4_4 contents as cell array
%        contents{get(hObject,'Value')} returns selected item from type_q_4_4


% --- Executes during object creation, after setting all properties.
function type_q_4_5_CreateFcn(hObject, eventdata, handles)
% hObject    handle to type_q_4_5 (see GCBO)
% eventdata  reserved - to be defined in a future version of MATLAB
% handles    empty - handles not created until after all CreateFcns called

% Hint: popupmenu controls usually have a white background on Windows.
%       See ISPC and COMPUTER.
if ispc
set(hObject,'BackgroundColor','white');
else
set(hObject,'BackgroundColor',get(0,'defaultUicontrolBackgroundColor'));
end

% --- Executes on selection change in type_q_4_5.
function type_q_4_5_Callback(hObject, eventdata, handles)
% hObject    handle to type_q_4_5 (see GCBO)
% eventdata  reserved - to be defined in a future version of MATLAB
% handles    structure with handles and user data (see GUIDATA)

% Hints: contents = get(hObject,'String') returns type_q_4_5 contents as cell array
```

```
1387 % contents{get(hObject,'Value')} returns selected item from type_q_4_5
1388
1389
1390 % --- Executes during object creation, after setting all properties.
1391 function type_q_4_6_CreateFcn(hObject, eventdata, handles)
1392 % hObject    handle to type_q_4_6 (see GCBO)
1393 % eventdata  reserved - to be defined in a future version of MATLAB
1394 % handles    empty - handles not created until after all CreateFcns called
1395
1396 % Hint: popupmenu controls usually have a white background on Windows.
1397 %       See ISPC and COMPUTER.
1398 if ispc
1399     set(hObject,'BackgroundColor','white');
1400 else
1401     set(hObject,'BackgroundColor',get(0,'defaultUicontrolBackgroundColor'));
1402 end
1403
1404 % --- Executes on selection change in type_q_4_6.
1405 function type_q_4_6_Callback(hObject, eventdata, handles)
1406 % hObject    handle to type_q_4_6 (see GCBO)
1407 % eventdata  reserved - to be defined in a future version of MATLAB
1408 % handles    structure with handles and user data (see GUIDATA)
1409
1410 % Hints: contents = get(hObject,'String') returns type_q_4_6 contents as cell array
1411 %        contents{get(hObject,'Value')} returns selected item from type_q_4_6
1412
1413
1414 % --- Executes during object creation, after setting all properties.
1415 function type_q_4_7_CreateFcn(hObject, eventdata, handles)
1416 % hObject    handle to type_q_4_7 (see GCBO)
1417 % eventdata  reserved - to be defined in a future version of MATLAB
1418 % handles    empty - handles not created until after all CreateFcns called
1419
1420 % Hint: popupmenu controls usually have a white background on Windows.
1421 %       See ISPC and COMPUTER.
1422 if ispc
1423     set(hObject,'BackgroundColor','white');
1424 else
1425     set(hObject,'BackgroundColor',get(0,'defaultUicontrolBackgroundColor'));
1426 end
1427
1428 % --- Executes on selection change in type_q_4_7.
1429 function type_q_4_7_Callback(hObject, eventdata, handles)
1430 % hObject    handle to type_q_4_7 (see GCBO)
1431 % eventdata  reserved - to be defined in a future version of MATLAB
1432 % handles    structure with handles and user data (see GUIDATA)
1433
1434 % Hints: contents = get(hObject,'String') returns type_q_4_7 contents as cell array
1435 %        contents{get(hObject,'Value')} returns selected item from type_q_4_7
1436
1437
1438 % --- Executes during object creation, after setting all properties.
1439 function type_q_4_8_CreateFcn(hObject, eventdata, handles)
1440 % hObject    handle to type_q_4_8 (see GCBO)
1441 % eventdata  reserved - to be defined in a future version of MATLAB
1442 % handles    empty - handles not created until after all CreateFcns called
1443
1444 % Hint: popupmenu controls usually have a white background on Windows.
1445 %       See ISPC and COMPUTER.
1446 if ispc
```

```
1447 set(hObject,'BackgroundColor','white');
1448 else
1449 set(hObject,'BackgroundColor',get(0,'defaultUicontrolBackgroundColor'));
1450 end
1451
1452 % --- Executes on selection change in type_q_4_8.
1453 function type_q_4_8_Callback(hObject, eventdata, handles)
1454 % hObject handle to type_q_4_8 (see GCBO)
1455 % eventdata reserved - to be defined in a future version of MATLAB
1456 % handles structure with handles and user data (see GUIDATA)
1457
1458 % Hints: contents = get(hObject,'String') returns type_q_4_8 contents as cell array
1459 % contents{get(hObject,'Value')} returns selected item from type_q_4_8
1460
1461
1462 % --- Executes during object creation, after setting all properties.
1463 function type_q_4_9_CreateFcn(hObject, eventdata, handles)
1464 % hObject handle to type_q_4_9 (see GCBO)
1465 % eventdata reserved - to be defined in a future version of MATLAB
1466 % handles empty - handles not created until after all CreateFcns called
1467
1468 % Hint: popupmenu controls usually have a white background on Windows.
1469 % See ISPC and COMPUTER.
1470 if ispc
1471 set(hObject,'BackgroundColor','white');
1472 else
1473 set(hObject,'BackgroundColor',get(0,'defaultUicontrolBackgroundColor'));
1474 end
1475
1476 % --- Executes on selection change in type_q_4_9.
1477 function type_q_4_9_Callback(hObject, eventdata, handles)
1478 % hObject handle to type_q_4_9 (see GCBO)
1479 % eventdata reserved - to be defined in a future version of MATLAB
1480 % handles structure with handles and user data (see GUIDATA)
1481
1482 % Hints: contents = get(hObject,'String') returns type_q_4_9 contents as cell array
1483 % contents{get(hObject,'Value')} returns selected item from type_q_4_9
1484
1485
1486 % --- Executes during object creation, after setting all properties.
1487 function type_q_4_10_CreateFcn(hObject, eventdata, handles)
1488 % hObject handle to type_q_4_10 (see GCBO)
1489 % eventdata reserved - to be defined in a future version of MATLAB
1490 % handles empty - handles not created until after all CreateFcns called
1491
1492 % Hint: popupmenu controls usually have a white background on Windows.
1493 % See ISPC and COMPUTER.
1494 if ispc
1495 set(hObject,'BackgroundColor','white');
1496 else
1497 set(hObject,'BackgroundColor',get(0,'defaultUicontrolBackgroundColor'));
1498 end
1499
1500 % --- Executes on selection change in type_q_4_10.
1501 function type_q_4_10_Callback(hObject, eventdata, handles)
1502 % hObject handle to type_q_4_10 (see GCBO)
1503 % eventdata reserved - to be defined in a future version of MATLAB
1504 % handles structure with handles and user data (see GUIDATA)
1505
```

```
1506 % Hints: contents = get(hObject,'String') returns type_q_4_10 contents as cell
array
1507 % contents{get(hObject,'Value')} returns selected item from type_q_4_10
1508
1509
1510 % --- Executes during object creation, after setting all properties.
1511 function type_q_4_11_CreateFcn(hObject, eventdata, handles)
1512 % hObject handle to type_q_4_11 (see GCBO)
1513 % eventdata reserved - to be defined in a future version of MATLAB
1514 % handles empty - handles not created until after all CreateFcns called
1515
1516 % Hint: popupmenu controls usually have a white background on Windows.
1517 % See ISPC and COMPUTER.
1518 if ispc
1519 set(hObject,'BackgroundColor','white');
1520 else
1521 set(hObject,'BackgroundColor',get(0,'defaultUicontrolBackgroundColor'));
1522 end
1523
1524 % --- Executes on selection change in type_q_4_11.
1525 function type_q_4_11_Callback(hObject, eventdata, handles)
1526 % hObject handle to type_q_4_11 (see GCBO)
1527 % eventdata reserved - to be defined in a future version of MATLAB
1528 % handles structure with handles and user data (see GUIDATA)
1529
1530 % Hints: contents = get(hObject,'String') returns type_q_4_11 contents as cell
array
1531 % contents{get(hObject,'Value')} returns selected item from type_q_4_11
1532
1533
1534 % --- Executes during object creation, after setting all properties.
1535 function type_q_4_12_CreateFcn(hObject, eventdata, handles)
1536 % hObject handle to type_q_4_12 (see GCBO)
1537 % eventdata reserved - to be defined in a future version of MATLAB
1538 % handles empty - handles not created until after all CreateFcns called
1539
1540 % Hint: popupmenu controls usually have a white background on Windows.
1541 % See ISPC and COMPUTER.
1542 if ispc
1543 set(hObject,'BackgroundColor','white');
1544 else
1545 set(hObject,'BackgroundColor',get(0,'defaultUicontrolBackgroundColor'));
1546 end
1547
1548 % --- Executes on selection change in type_q_4_12.
1549 function type_q_4_12_Callback(hObject, eventdata, handles)
1550 % hObject handle to type_q_4_12 (see GCBO)
1551 % eventdata reserved - to be defined in a future version of MATLAB
1552 % handles structure with handles and user data (see GUIDATA)
1553
1554 % Hints: contents = get(hObject,'String') returns type_q_4_12 contents as cell
array
1555 % contents{get(hObject,'Value')} returns selected item from type_q_4_12
1556
1557
1558
1559 % --- Executes during object creation, after setting all properties.
1560 function type_q_3_1_CreateFcn(hObject, eventdata, handles)
1561 % hObject handle to type_q_3_1 (see GCBO)
1562 % eventdata reserved - to be defined in a future version of MATLAB
1563 % handles empty - handles not created until after all CreateFcns called
```

```matlab
% Hint: popupmenu controls usually have a white background on Windows.
%       See ISPC and COMPUTER.
if ispc
set(hObject,'BackgroundColor','white');
else
set(hObject,'BackgroundColor',get(0,'defaultUicontrolBackgroundColor'));
end

% --- Executes on selection change in type_q_3_1.
function type_q_3_1_Callback(hObject, eventdata, handles)
% hObject handle to type_q_3_1 (see GCBO)
% eventdata reserved - to be defined in a future version of MATLAB
% handles structure with handles and user data (see GUIDATA)

% Hints: contents = get(hObject,'String') returns type_q_3_1 contents as cell array
%        contents{get(hObject,'Value')} returns selected item from type_q_3_1


% --- Executes during object creation, after setting all properties.
function type_q_3_2_CreateFcn(hObject, eventdata, handles)
% hObject handle to type_q_3_2 (see GCBO)
% eventdata reserved - to be defined in a future version of MATLAB
% handles empty - handles not created until after all CreateFcns called

% Hint: popupmenu controls usually have a white background on Windows.
%       See ISPC and COMPUTER.
if ispc
set(hObject,'BackgroundColor','white');
else
set(hObject,'BackgroundColor',get(0,'defaultUicontrolBackgroundColor'));
end

% --- Executes on selection change in type_q_3_2.
function type_q_3_2_Callback(hObject, eventdata, handles)
% hObject handle to type_q_3_2 (see GCBO)
% eventdata reserved - to be defined in a future version of MATLAB
% handles structure with handles and user data (see GUIDATA)

% Hints: contents = get(hObject,'String') returns type_q_3_2 contents as cell array
%        contents{get(hObject,'Value')} returns selected item from type_q_3_2


% --- Executes during object creation, after setting all properties.
function type_q_3_3_CreateFcn(hObject, eventdata, handles)
% hObject handle to type_q_3_3 (see GCBO)
% eventdata reserved - to be defined in a future version of MATLAB
% handles empty - handles not created until after all CreateFcns called

% Hint: popupmenu controls usually have a white background on Windows.
%       See ISPC and COMPUTER.
if ispc
set(hObject,'BackgroundColor','white');
else
set(hObject,'BackgroundColor',get(0,'defaultUicontrolBackgroundColor'));
end

% --- Executes on selection change in type_q_3_3.
function type_q_3_3_Callback(hObject, eventdata, handles)
% hObject handle to type_q_3_3 (see GCBO)
```

```matlab
1624 % eventdata  reserved - to be defined in a future version of MATLAB
1625 % handles    structure with handles and user data (see GUIDATA)
1626
1627 % Hints: contents = get(hObject,'String') returns type_q_3_3 contents as cell array
1628 %        contents{get(hObject,'Value')} returns selected item from type_q_3_3
1629
1630
1631 % --- Executes during object creation, after setting all properties.
1632 function type_q_3_4_CreateFcn(hObject, eventdata, handles)
1633 % hObject    handle to type_q_3_4 (see GCBO)
1634 % eventdata  reserved - to be defined in a future version of MATLAB
1635 % handles    empty - handles not created until after all CreateFcns called
1636
1637 % Hint: popupmenu controls usually have a white background on Windows.
1638 %       See ISPC and COMPUTER.
1639 if ispc
1640     set(hObject,'BackgroundColor','white');
1641 else
1642     set(hObject,'BackgroundColor',get(0,'defaultUicontrolBackgroundColor'));
1643 end
1644
1645 % --- Executes on selection change in type_q_3_4.
1646 function type_q_3_4_Callback(hObject, eventdata, handles)
1647 % hObject    handle to type_q_3_4 (see GCBO)
1648 % eventdata  reserved - to be defined in a future version of MATLAB
1649 % handles    structure with handles and user data (see GUIDATA)
1650
1651 % Hints: contents = get(hObject,'String') returns type_q_3_4 contents as cell array
1652 %        contents{get(hObject,'Value')} returns selected item from type_q_3_4
1653
1654
1655 % --- Executes during object creation, after setting all properties.
1656 function type_q_3_5_CreateFcn(hObject, eventdata, handles)
1657 % hObject    handle to type_q_3_5 (see GCBO)
1658 % eventdata  reserved - to be defined in a future version of MATLAB
1659 % handles    empty - handles not created until after all CreateFcns called
1660
1661 % Hint: popupmenu controls usually have a white background on Windows.
1662 %       See ISPC and COMPUTER.
1663 if ispc
1664     set(hObject,'BackgroundColor','white');
1665 else
1666     set(hObject,'BackgroundColor',get(0,'defaultUicontrolBackgroundColor'));
1667 end
1668
1669 % --- Executes on selection change in type_q_3_5.
1670 function type_q_3_5_Callback(hObject, eventdata, handles)
1671 % hObject    handle to type_q_3_5 (see GCBO)
1672 % eventdata  reserved - to be defined in a future version of MATLAB
1673 % handles    structure with handles and user data (see GUIDATA)
1674
1675 % Hints: contents = get(hObject,'String') returns type_q_3_5 contents as cell array
1676 %        contents{get(hObject,'Value')} returns selected item from type_q_3_5
1677
1678
1679 % --- Executes during object creation, after setting all properties.
1680 function type_q_3_6_CreateFcn(hObject, eventdata, handles)
1681 % hObject    handle to type_q_3_6 (see GCBO)
```

```
% eventdata reserved - to be defined in a future version of MATLAB
% handles empty - handles not created until after all CreateFcns called

% Hint: popupmenu controls usually have a white background on Windows.
% See ISPC and COMPUTER.
if ispc
set(hObject,'BackgroundColor','white');
else
set(hObject,'BackgroundColor',get(0,'defaultUicontrolBackgroundColor'));
end

% --- Executes on selection change in type_q_3_6.
function type_q_3_6_Callback(hObject, eventdata, handles)
% hObject handle to type_q_3_6 (see GCBO)
% eventdata reserved - to be defined in a future version of MATLAB
% handles structure with handles and user data (see GUIDATA)

% Hints: contents = get(hObject,'String') returns type_q_3_6 contents as cell array
% contents{get(hObject,'Value')} returns selected item from type_q_3_6


% --- Executes during object creation, after setting all properties.
function type_q_3_7_CreateFcn(hObject, eventdata, handles)
% hObject handle to type_q_3_7 (see GCBO)
% eventdata reserved - to be defined in a future version of MATLAB
% handles empty - handles not created until after all CreateFcns called

% Hint: popupmenu controls usually have a white background on Windows.
% See ISPC and COMPUTER.
if ispc
set(hObject,'BackgroundColor','white');
else
set(hObject,'BackgroundColor',get(0,'defaultUicontrolBackgroundColor'));
end

% --- Executes on selection change in type_q_3_7.
function type_q_3_7_Callback(hObject, eventdata, handles)
% hObject handle to type_q_3_7 (see GCBO)
% eventdata reserved - to be defined in a future version of MATLAB
% handles structure with handles and user data (see GUIDATA)

% Hints: contents = get(hObject,'String') returns type_q_3_7 contents as cell array
% contents{get(hObject,'Value')} returns selected item from type_q_3_7


% --- Executes during object creation, after setting all properties.
function type_q_3_8_CreateFcn(hObject, eventdata, handles)
% hObject handle to type_q_3_8 (see GCBO)
% eventdata reserved - to be defined in a future version of MATLAB
% handles empty - handles not created until after all CreateFcns called

% Hint: popupmenu controls usually have a white background on Windows.
% See ISPC and COMPUTER.
if ispc
set(hObject,'BackgroundColor','white');
else
set(hObject,'BackgroundColor',get(0,'defaultUicontrolBackgroundColor'));
end

% --- Executes on selection change in type_q_3_8.
```

```
1742 function type_q_3_8_Callback(hObject, eventdata, handles)
1743 % hObject handle to type_q_3_8 (see GCBO)
1744 % eventdata reserved - to be defined in a future version of MATLAB
1745 % handles structure with handles and user data (see GUIDATA)
1746
1747 % Hints: contents = get(hObject,'String') returns type_q_3_8 contents as cell array
1748 %        contents{get(hObject,'Value')} returns selected item from type_q_3_8
1749
1750
1751 % --- Executes during object creation, after setting all properties.
1752 function type_q_3_9_CreateFcn(hObject, eventdata, handles)
1753 % hObject handle to type_q_3_9 (see GCBO)
1754 % eventdata reserved - to be defined in a future version of MATLAB
1755 % handles empty - handles not created until after all CreateFcns called
1756
1757 % Hint: popupmenu controls usually have a white background on Windows.
1758 %       See ISPC and COMPUTER.
1759 if ispc
1760 set(hObject,'BackgroundColor','white');
1761 else
1762 set(hObject,'BackgroundColor',get(0,'defaultUicontrolBackgroundColor'));
1763 end
1764
1765 % --- Executes on selection change in type_q_3_9.
1766 function type_q_3_9_Callback(hObject, eventdata, handles)
1767 % hObject handle to type_q_3_9 (see GCBO)
1768 % eventdata reserved - to be defined in a future version of MATLAB
1769 % handles structure with handles and user data (see GUIDATA)
1770
1771 % Hints: contents = get(hObject,'String') returns type_q_3_9 contents as cell array
1772 %        contents{get(hObject,'Value')} returns selected item from type_q_3_9
1773
1774
1775 % --- Executes during object creation, after setting all properties.
1776 function type_q_3_10_CreateFcn(hObject, eventdata, handles)
1777 % hObject handle to type_q_3_10 (see GCBO)
1778 % eventdata reserved - to be defined in a future version of MATLAB
1779 % handles empty - handles not created until after all CreateFcns called
1780
1781 % Hint: popupmenu controls usually have a white background on Windows.
1782 %       See ISPC and COMPUTER.
1783 if ispc
1784 set(hObject,'BackgroundColor','white');
1785 else
1786 set(hObject,'BackgroundColor',get(0,'defaultUicontrolBackgroundColor'));
1787 end
1788
1789 % --- Executes on selection change in type_q_3_10.
1790 function type_q_3_10_Callback(hObject, eventdata, handles)
1791 % hObject handle to type_q_3_10 (see GCBO)
1792 % eventdata reserved - to be defined in a future version of MATLAB
1793 % handles structure with handles and user data (see GUIDATA)
1794
1795 % Hints: contents = get(hObject,'String') returns type_q_3_10 contents as cell array
1796 %        contents{get(hObject,'Value')} returns selected item from type_q_3_10
1797
1798
1799 % --- Executes during object creation, after setting all properties.
```

```matlab
function type_q_3_11_CreateFcn(hObject, eventdata, handles)
% hObject handle to type_q_3_11 (see GCBO)
% eventdata reserved - to be defined in a future version of MATLAB
% handles empty - handles not created until after all CreateFcns called

% Hint: popupmenu controls usually have a white background on Windows.
% See ISPC and COMPUTER.
if ispc
set(hObject,'BackgroundColor','white');
else
set(hObject,'BackgroundColor',get(0,'defaultUicontrolBackgroundColor'));
end

% --- Executes on selection change in type_q_3_11.
function type_q_3_11_Callback(hObject, eventdata, handles)
% hObject handle to type_q_3_11 (see GCBO)
% eventdata reserved - to be defined in a future version of MATLAB
% handles structure with handles and user data (see GUIDATA)

% Hints: contents = get(hObject,'String') returns type_q_3_11 contents as cell array
% contents{get(hObject,'Value')} returns selected item from type_q_3_11


% --- Executes during object creation, after setting all properties.
function type_q_3_12_CreateFcn(hObject, eventdata, handles)
% hObject handle to type_q_3_12 (see GCBO)
% eventdata reserved - to be defined in a future version of MATLAB
% handles empty - handles not created until after all CreateFcns called

% Hint: popupmenu controls usually have a white background on Windows.
% See ISPC and COMPUTER.
if ispc
set(hObject,'BackgroundColor','white');
else
set(hObject,'BackgroundColor',get(0,'defaultUicontrolBackgroundColor'));
end

% --- Executes on selection change in type_q_3_12.
function type_q_3_12_Callback(hObject, eventdata, handles)
% hObject handle to type_q_3_12 (see GCBO)
% eventdata reserved - to be defined in a future version of MATLAB
% handles structure with handles and user data (see GUIDATA)

% Hints: contents = get(hObject,'String') returns type_q_3_12 contents as cell array
% contents{get(hObject,'Value')} returns selected item from type_q_3_12



% --- Executes during object creation, after setting all properties.
function next_ac_type_CreateFcn(hObject, eventdata, handles)
% hObject handle to next_ac_type (see GCBO)
% eventdata reserved - to be defined in a future version of MATLAB
% handles empty - handles not created until after all CreateFcns called

% Hint: popupmenu controls usually have a white background on Windows.
% See ISPC and COMPUTER.
if ispc
set(hObject,'BackgroundColor','white');
else
set(hObject,'BackgroundColor',get(0,'defaultUicontrolBackgroundColor'));
```

```
1860 end
1861
1862
1863 % --- Executes on selection change in next_ac_type.
1864 function next_ac_type_Callback(hObject, eventdata, handles)
1865 % hObject  handle to next_ac_type (see GCBO)
1866 % eventdata  reserved - to be defined in a future version of MATLAB
1867 % handles    structure with handles and user data (see GUIDATA)
1868
1869 % Hints: contents = get(hObject,'String') returns next_ac_type contents
as cell
array
1870 %        contents{get(hObject,'Value')} returns selected item from next_ac_type
1871
1872
1873
1874 %--------------------------------------------------------------------------
1875 % Push bottom to run the simulation
1876 %--------------------------------------------------------------------------
1877
1878 % --- Executes on button press in run_simul.
1879 function run_simul_Callback(hObject, eventdata, handles)
1880 % hObject  handle to run_simul (see GCBO)
1881 % eventdata  reserved - to be defined in a future version of MATLAB
1882 % handles    structure with handles and user data (see GUIDATA)
1883
1884 % debug
1885 bug_check = 0;
1886
1887 % The parameters for the simulation:
1888 %
1889 % The type of distribution: 1 for Gaussian, 2 for exponential,
1890 % 3 for exponential @ the stations and Gaussian @ queue and taxi, and
1891 % 4 for exponential @ the stations and lognormal @ queue and taxi is
1892 % normal
1893 dst = 4;
1894
1895 % Input the simulation length
1896 % simul = str2double(get(handles.simul_len,'String'));
1897 simul = 1000;
1898
1899 % Load configuration data
1900 load dtw_deice_data.mat;
1901
1902 % Initialize display options,
1903 % set(handles.text_win_4_sys, 'String', []);
1904 % set(handles.text_win_4_ci1, 'String', []);
1905 % set(handles.text_win_4_ci2, 'String', []);
1906 % set(handles.text_win_3_sys, 'String', []);
1907 % set(handles.text_win_3_ci1, 'String', []);
1908 % set(handles.text_win_3_ci2, 'String', []);
1909 axes(handles.fig_result);
1910 refresh(handles.dtw_deice_gui);
1911 set(handles.fig_result, 'Visible', 'off');
1912 set(handles.fig_cover, 'Visible', 'on');
1913 set(handles.fig_cover, 'String', {' '; ' '; ' '; ' '; ' '; ' '; ' ';
'Simulating'});
1914 drawnow;
1915
1916
1917 % The confidence interval on the plot and the corresponding percentiles
1918 conf = 95;
```

```
pertile_low = (100 - conf) / 2;
pertile_high = 100 - pertile_low;

% The name of the pads:
deice_pad(1).name = '4R';
deice_pad(2).name = '3L';

% The size of the pads:
deice_pad(1).num_pos = 7;
deice_pad(2).num_pos = 7;

% Initialize the event parameters
event(1).name = 'deice';
event(2).name = 'taxi';

% The number of type of events in this simulation
event_num = 2;

% The deicing time characterization
% The rows identify the AC type, corresponding to the numerical code of
% AC type below. The columns identify the event type.

% Input the various parameters for the simulation
%
% Input the snow event type:
snow_type = get(handles.type_snow,'Value');

% The average deice time
load dtw_deice_data.mat;
event(1).m = pad_time;
event(1).m(1, snow_type) = str2double(get(handles.time_rj, 'String'));
event(1).m(2, snow_type) = str2double(get(handles.time_dc9, 'String'));
event(1).m(3, snow_type) = str2double(get(handles.time_a320, 'String'));
event(1).m(4, snow_type) = str2double(get(handles.time_b757, 'String'));
event(1).m(5, snow_type) = str2double(get(handles.time_dc10, 'String'));
event(1).m(6, snow_type) = str2double(get(handles.time_b747, 'String'));

% Error check: all pad time should be positive
temp = find(event(1).m <= 0);
if ~isempty(temp)
msgbox('Error: Must be Positive Values for the Pad Times', 'Pad Time Error',
'error');
return;
end

% The standard deviation of pad time
event(1).sd = [3.93, 1.29, 6.18, 0.13, 4.89, 6.18, 6.18; ...
4.44, 2.61, 7.16, 6.94, 7.95, 7.16, 7.16; ...
4.16, 3.83, 7.91, 9.16, 8.50, 7.91, 7.91; ...
4.75, 2.70, 7.95, 10.89, 8.40, 7.95, 7.95; ...
0.5, 0.5, 0.5, 0.5, 0.5, 0.5, 0.5; ...
0.5, 0.5, 0.5, 0.5, 0.5, 0.5, 0.5];
% debug
% deice_sd = 0 * ones(6, 7);

% Convert mean and standard deviation of a lognormal distribution to log
% mean and log standard deviation.
if (dst == 4)
[event(1).logm, event(1).logsd] = m_logm(event(1).m, event(1).sd);
end

% The average taxi time
event(2).m = ones(6,7);
```

```
1981 % The standard deviation of taxi time
1982 event(2).sd = ones(6,7) / 6;
1983
1984 % The size restriction of each station
1985 %
1986 % The size restriction at 4R pad
1987 deice_pad(1).size_res = [length(get(handles.type_s_4_1, 'String')); length(get(handles.type_s_4_2, 'String')); ...
1988 length(get(handles.type_s_4_3, 'String')); length(get(handles.type_s_4_4, 'String')); ...
1989 length(get(handles.type_s_4_5, 'String')); length(get(handles.type_s_4_6, 'String')); ...
1990 length(get(handles.type_s_4_7, 'String'))] - ones(7,1);
1991
1992
1993 % The size restriction at 3L pad
1994 deice_pad(2).size_res = [length(get(handles.type_s_3_1, 'String')); length(get(handles.type_s_3_2, 'String')); ...
1995 length(get(handles.type_s_3_3, 'String')); length(get(handles.type_s_3_4, 'String')); ...
1996 length(get(handles.type_s_3_5, 'String')); length(get(handles.type_s_3_6, 'String')); ...
1997 length(get(handles.type_s_3_7, 'String'))] - ones(7,1);
1998
1999
2000
2001 % Input the state of the server
2002 %
2003 % Input the AC type of the outbound AC
2004 type_next = 7 - get(handles.next_ac_type,'Value');
2005
2006
2007 % Input the state of 4R deice pad
2008 for pos = 1:deice_pad(1).num_pos,
2009 % Inpute the AC type at the server
2010 eval(['deice_pad(1).server.ac_type(', num2str(pos), ') = length (ac_list_s_4_', num2str(pos), ') - '...
2011 'get(handles.type_s_4_', num2str(pos), ',''Value'');']);
2012 % Input the time elapsed
2013 eval(['deice_pad(1).server.elap_time(', num2str(pos), ') = str2double(get(handles.time_s_4_', num2str(pos), ', ...
2014 ',''String''));']);
2015 % Obtain the mean time and SD for the service time
2016 % if the station is closed
2017 if eval(['get(handles.checkbox_4_', num2str(pos), ', ''Value'') == 0'])
2018 deice_pad(1).server.m(pos) = NaN;
2019 deice_pad(1).server.sd(pos) = NaN;
2020 deice_pad(1).server.ac_type(pos) = NaN;
2021 deice_pad(1).server.elap_time(pos) = NaN;
2022 % if the station is empty
2023 elseif deice_pad(1).server.ac_type(pos) == 0
2024 deice_pad(1).server.m(pos) = 0;
2025 deice_pad(1).server.sd(pos) = 0;
2026 % if the station is not empty
2027 else
2028 deice_pad(1).server.m(pos) = event(1).m(deice_pad(1).server.ac_type(pos), snow_type) - deice_pad(1).server.elap_time(pos);
2029 deice_pad(1).server.sd(pos) = event(1).sd(deice_pad(1).server.ac_type(pos), snow_type);
2030 % debug
2031 % deice_pad(1).server.sd(pos) = 0;
2032 end
```

```
2033 end
2034
2035
2036 % Input the state of the queue for 4R
2037 %
2038 % Inpute the AC type at the queue
2039 for count = 1:12,
2040 eval(['deice_pad(1).queue.ac_type(', num2str(count), ') = length(ac_list_q_4)
+ 1 - get(handles.type_q_4_', num2str(count), ...
2041 ',''Value'');']);
2042 end
2043 % Remove any empty slots in the queue
2044 temp = find(deice_pad(1).queue.ac_type == length(ac_list_q_4));
2045 deice_pad(1).queue.ac_type(temp) = [];
2046 % Insert the outbound AC
2047 deice_pad(1).queue.ac_type = [deice_pad(1).queue.ac_type type_next];
2048
2049
2050
2051 % Input 3L deice pad
2052 % Inpute the AC type at the server
2053 for pos = 1:deice_pad(2).num_pos,
2054 % Inpute the AC type at the server
2055 eval(['deice_pad(2).server.ac_type(', num2str(pos), ') = length
(ac_list_s_3_', num2str(pos), ') - '...
2056 'get(handles.type_s_3_', num2str(pos), ',''Value'');']);
2057 % Input the time elapsed
2058 eval(['deice_pad(2).server.elap_time(', num2str(pos), ') = str2double(get
(handles.time_s_3_', num2str(pos), ...
2059 ',''String'');']);
2060 % Obtain the mean time and SD for the service time
2061 % if the station is closed
2062 if eval(['get(handles.checkbox_3_', num2str(pos), ', ''Value'') == 0'])
2063 deice_pad(2).server.m(pos) = NaN;
2064 deice_pad(2).server.sd(pos) = NaN;
2065 % if the station is empty
2066 elseif deice_pad(2).server.ac_type(pos) == 0
2067 deice_pad(2).server.m(pos) = 0;
2068 deice_pad(2).server.sd(pos) = 0;
2069 % if the station is not empty
2070 else
2071 deice_pad(2).server.m(pos) = event(1).m(deice_pad(2).server.ac_type
(pos), snow_type) - deice_pad(2).server.elap_time(pos);
2072 deice_pad(2).server.sd(pos) = event(1).sd(deice_pad(2).server.ac_type
(pos), snow_type);
2073 % debug
2074 % deice_pad(2).server.sd(pos) = 0;
2075 end
2076 end
2077
2078
2079 % Input the state of the queue 3L
2080 %
2081 % Inpute the AC type at the queue
2082 for count = 1:12,
2083 eval(['deice_pad(2).queue.ac_type(', num2str(count), ') = length(ac_list_q_3)
+ 1 - get(handles.type_q_3_', num2str(count), ...
2084 ',''Value'');']);
2085 end
2086 % Remove any empty slots in the queue
2087 temp = find(deice_pad(2).queue.ac_type == length(ac_list_q_3));
2088 deice_pad(2).queue.ac_type(temp) = [];
```

```
2089 % Insert the outbound AC
2090 deice_pad(2).queue.ac_type = [deice_pad(2).queue.ac_type type_next];
2091
2092
2093 fprintf('\n-------------------------------------------------------------------\n');
2094
2095 for pad = 1:2,
2096
2097 % debug
2098 if bug_check
2099 pad
2100 end
2101
2102 % Initialize pad restriction
2103 pad_res = deice_pad(pad).size_res;
2104
2105 % Check to see if the AC is too big for 3L
2106 if ((pad == 2) && (type_next > max(pad_res)))
2107 % m(pad) = er_high(1) + 60;
2108 % sd(pad) = 0;
2109 % er_low(pad) = er_high(1) + 60;
2110 % er_high(pad) = er_high(1) + 60;
2111 m(pad) = NaN;
2112 sd(pad) = NaN;
2113 er_low(pad) = NaN;
2114 er_high(pad) = NaN;
2115 fprintf('\n*************************************************************\n');
2116 fprintf('\n Warning: the outbound aircraft is too big to fit in pad %s\n', deice_pad(pad).name);
2117 fprintf('\n*************************************************************\n');
2118 continue;
2119 end
2120
2121
2122 % The queuing simulation
2123 % Initialize the master clock
2124 deice_pad(pad).t = zeros(simul, 1);
2125 for count = 1:simul,
2126
2127 % Initialize the simulation
2128 % Initialize the event durations
2129 if (dst == 1)
2130 event(1).t = normrnd(deice_pad(pad).server.m, deice_pad(pad).server.sd);
2131 elseif (dst == 2 || dst == 3 || dst == 4)
2132 event(1).t = exprnd(deice_pad(pad).server.m);
2133 % debug
2134 if bug_check
2135 fprintf('\nevent(1).t\n');
2136 event(1).t
2137 end
2138
2139 end
2140 event(2).t = [];
2141 % Initialize the server_ac and q_ac
2142 event(1).ac = deice_pad(pad).server.ac_type;
2143 % debug
2144 if bug_check
2145 fprintf('\nevent(1).ac\n');
2146 event(1).ac
```

```
2147 end
2148
2149 event(2).ac = [];
2150 event(2).station = [];
2151 q_ac = deice_pad(pad).queue.ac_type;
2152
2153 % Initialize the flag for the last AC in queue
2154 last_ac_flag = isempty(q_ac);
2155 % Initialize the minimum of each event type
2156 for count_event = 1 : event_num,
2157 if (isempty(event(count_event).t))
2158 minimum(count_event) = NaN;
2159 else
2160 minimum(count_event) = min(event(count_event).t);
2161 end
2162 end
2163
2164 % Check for events and update the state
2165 while (~isempty(q_ac) || last_ac_flag),
2166 % Identify the next event
2167 % If this is the last AC from queue is in server, end the
2168 % simulation
2169 if (last_ac_flag == 1)
2170 if (dst == 1 || dst == 3)
2171 deice_pad(pad).t(count) = deice_pad(pad).t(count) + ...
2172 normrnd(event(1).m(event(2).ac(end), snow_type), event(1).sd(event(2).ac(end), snow_type)) + ...
2173 normrnd(event(2).m(event(2).ac(end), snow_type), event(2).sd(event(2).ac(end), snow_type));
2174 elseif (dst == 2)
2175 deice_pad(pad).t(count) = deice_pad(pad).t(count) + ...
2176 exprnd(event(1).m(event(2).ac(end), snow_type)) + ...
2177 exprnd(event(2).m(event(2).ac(end), snow_type));
2178 elseif (dst == 4)
2179 deice_pad(pad).t(count) = deice_pad(pad).t(count) + ...
2180 lognrnd(event(1).logm(event(2).ac(end), snow_type), event(1).logsd(event(2).ac(end), snow_type));
2181 end
2182 break
2183 else
2184 % Identify event
2185 [temp, event_type] = min(minimum);
2186 % Update the master clock
2187 deice_pad(pad).t(count) = deice_pad(pad).t(count) + temp;
2188 % debug
2189 if bug_check
2190 temp
2191 event_type
2192 fprintf('\ndeice_pad(pad).t(count)\n');
2193 deice_pad(pad).t
2194 end
2195
2196 % Update the event time
2197 for count_event = 1 : event_num,
2198 event(count_event).t = event(count_event).t - temp;
2199 end
2200 % Update the state
2201 if (event_type == 1)
2202 % ID the station that just emptied
2203 station = find(event(1).t == 0) ;
2204 % Reset the event time for the empty station
2205 event(1).t(station) = NaN;
2206 % Reset the status of the empty station to empty
2207 event(1).ac(station) = 0;
```

```
2208 % ID empty stations
2209 server_empty = find(event(1).ac == 0);
2210
2211 % debug
2212 if bug_check
2213 fprintf('\nevent(1).ac\n');
2214 event(1).ac
2215 fprintf('\nevent(1).t\n');
2216 event(1).t
2217 fprintf('\nevent(2).ac\n');
2218 event(2).ac
2219 fprintf('\nevent(2).t\n');
2220 event(2).t
2221 fprintf('\nevent(2).station\n');
2222 event(2).station
2223 q_ac
2224 server_empty
2225 last_ac_flag
2226 end % end: if bug_check
2227
2228 % Move the first AC in queue into the taxi position if it fits the
2229 % one of the opening stations
2230 while (nnz(q_ac(1) <= deice_pad(pad).size_res(server_empty))) || ...
2231 ((widebody_flag) && (q_ac(1) >= 5) && (length(server_empty) > 1) && (~isempty(find(diff(server_empty) == 1)))),
2232
2233 % Check to see if the first AC in queue fits into one of the emptied
2234 % stations
2235 if (nnz(q_ac(1) <= deice_pad(pad).size_res(server_empty)))
2236 % Get the largest empty station number
2237 [val, ind] = max(server_empty(q_ac(1) <= deice_pad(pad).size_res(server_empty)));
2238 if (dst == 1 || dst == 3 || dst == 4)
2239 event(2).t = [event(2).t, normrnd(event(2).m(q_ac(1), snow_type), event(2).sd(q_ac(1), snow_type))];
2240 elseif (dst == 2)
2241 event(2).t = [event(2).t, exprnd(event(2).m(q_ac(1), snow_type))];
2242 end
2243 event(2).ac = [event(2).ac, q_ac(1)];
2244 event(2).station = [event(2).station val];
2245 event(1).ac(val) = q_ac(1);
2246 q_ac(1) = [];
2247 server_empty(ind) = [];
2248 last_ac_flag = isempty(q_ac);
2249 % If there is no airplane in queue, break this loop of
2250 % checking for empty spots.
2251 if (last_ac_flag)
2252 break
2253 end
2254 server_empty = find(event(1).ac == 0);
2255
2256 % debug
2257 if bug_check
2258 fprintf('\nevent(1).ac\n');
2259 event(1).ac
2260 fprintf('\nevent(1).t\n');
2261 event(1).t
2262 fprintf('\nevent(2).ac\n');
```

```
2263 event(2).ac
2264 fprintf('\nevent(2).t\n');
2265 event(2).t
2266 fprintf('\nevent(2).station\n');
2267 event(2).station
2268 q_ac
2269 server_empty
2270 last_ac_flag
2271 end % end: if bug_check
2272 end % end: if (nnz(q_ac(1) <= deice_pad(pad).size_res(server_empty)))
2273
2274 % Speical rules: alow widebody to take up two adjacent
2275 % empty slots.
2276 if (widebody_flag) && (q_ac(1) >= 5) && (length(server_empty) > 1) && (~isempty(find(diff(server_empty) == 1)))
2277 ind = max(find(diff(server_empty) == 1));
2278 val = server_empty(ind);
2279 if (dst == 1 || dst == 3 || dst == 4)
2280 event(2).t = [event(2).t, normrnd(event(2).m(q_ac(1), snow_type), event(2).sd(q_ac(1), snow_type))];
2281 elseif (dst == 2)
2282 event(2).t = [event(2).t, exprnd(event(2).m(q_ac(1), snow_type))];
2283 end
2284 event(2).ac = [event(2).ac, q_ac(1)];
2285 event(2).station = [event(2).station val];
2286 event(1).ac(val) = q_ac(1);
2287 event(1).ac(val + 1) = q_ac(1);
2288 q_ac(1) = [];
2289 server_empty(ind) = [];
2290 last_ac_flag = isempty(q_ac);
2291 % If there is no airplane in queue, break this loop of
2292 % checking for empty spots.
2293 if (last_ac_flag)
2294 break
2295 end
2296 server_empty = find(event(1).ac == 0);
2297
2298 % debug
2299 if bug_check
2300 fprintf('\nevent(1).ac\n');
2301 event(1).ac
2302 fprintf('\nevent(1).t\n');
2303 event(1).t
2304 fprintf('\nevent(2).ac\n');
2305 event(2).ac
2306 fprintf('\nevent(2).t\n');
2307 event(2).t
2308 fprintf('\nevent(2).station\n');
2309 event(2).station
2310 q_ac
2311 server_empty
2312 last_ac_flag
2313 end % end: if bug_check
2314 end % end:(widebody_flag) && (q_ac(1) >= 5) && (length(server_empty) > 1) && (~isempty(find(diff(server_empty) == 1)))
2315 end % end: (nnz(q_ac(1) <= deice_pad(pad).size_res(server_empty))) || ...
2316
2317 elseif (event_type == 2)
2318 ind = max(find(event(2).t == 0));
2319 event(1).ac(event(2).station(ind)) = event(2).ac(ind);
```

```
2320 if (dst == 1 || dst == 3)
2321 event(1).t(event(2).station(ind)) = normrnd(event(1).m(event(2).ac(ind), snow_type), ...
2322 event(1).sd(event(2).ac(ind), snow_type));
2323 % Special rule: alow widebody to take up two adjucent
2324 % empty slots.
2325 if event(1).ac(event(2).station(ind)) > pad_res(event(2).station(ind))
2326 event(1).t(event(2).station(ind) + 1) = event(1).t(event(2).station(ind));
2327 end
2328
2329 elseif (dst == 4)
2330 event(1).t(event(2).station(ind)) = lognrnd(event(1).logm(event(2).ac(ind), snow_type), ...
2331 event(1).logsd(event(2).ac(ind), snow_type));
2332 % Special rule: alow widebody to take up two adjucent
2333 % empty slots.
2334
2335 % debug
2336 if bug_check
2337 fprintf('\nind')
2338 ind
2339 fprintf('\nevent(1).ac(event(2).station(ind))');
2340 event(1).ac(event(2).station(ind))
2341 fprintf('pad_res(event(2).station(ind))')
2342 pad_res(event(2).station(ind))
2343 end
2344
2345 if event(1).ac(event(2).station(ind)) > pad_res(event(2).station(ind))
2346 event(1).t(event(2).station(ind) + 1) = event(1).t(event(2).station(ind));
2347 end
2348
2349 elseif (dst == 2)
2350 event(1).t(event(2).station(ind)) = exprnd(event(1).m(event(2).ac(ind), snow_type));
2351 % Special rule: alow widebody to take up two adjucent
2352 % empty slots.
2353 if event(1).ac(event(2).station(ind)) > pad_res(event(2).station(ind))
2354 event(1).t(event(2).station(ind) + 1) = event(1).t(event(2).station(ind));
2355 end
2356 end
2357 event(2).t(ind) = [];
2358 event(2).ac(ind) = [];
2359 event(2).station(ind) = [];
2360
2361 % debug
2362 if bug_check
2363 fprintf('\nevent(1).ac\n');
2364 event(1).ac
2365 fprintf('\nevent(1).t\n');
2366 event(1).t
2367 fprintf('\nevent(2).ac\n');
2368 event(2).ac
2369 fprintf('\nevent(2).t\n');
2370 event(2).t
2371 fprintf('\nevent(2).station\n');
2372 event(2).station
2373 q_ac
2374 server_empty
```

```
2375     last_ac_flag
2376   end % end: if bug_check
2377
2378 end
2379
2380 % Update minimum
2381 for count_event = 1 : event_num,
2382   if (isempty(event(count_event).t))
2383     minimum(count_event) = NaN;
2384   else
2385     minimum(count_event) = min(event(count_event).t);
2386   end
2387 end % end: for count_event = 1 : event_num,
2388 end % end: if (last_ac_flag)
2389 end % end: while
2390 end % end: for count = 1:simul,
2391 % Data analysis
2392 m(pad) = mean(deice_pad(pad).t);
2393 sd(pad) = std(deice_pad(pad).t);
2394 er_low(pad) = prctile(deice_pad(pad).t, pertile_low);
2395 er_high(pad) = prctile(deice_pad(pad).t, pertile_high);
2396 % Display result
2397 fprintf('For deice pad %s: \n', deice_pad(pad).name);
2398 fprintf('The average system time is %3.2f min.\n', m(pad));
2399 fprintf('The 95 percent confidence interval is from %3.2f min. to %3.2f min.
\n\n', er_low(pad), er_high(pad));
2400
2401 end
2402
2403
2404 % Initialize the display result for 4R
2405 % set(handles.text_win_4, 'Visible', 'on');
2406 % set(handles.text_win_4_sys, 'Visible', 'on');
2407 % set(handles.text_win_4_ci1, 'Visible', 'on');
2408 % set(handles.text_win_4_ci2, 'Visible', 'on');
2409 % Check to see if the AC is too big for 3L
2410 % if type_next > max(pad_res)
2411 %   fprintf
('\n************************************************************\n');
2412 %   fprintf('\nWarning: the incoming aircraft is too big to fit in pad %s
\n',
deice_pad(pad).name);
2413 %   fprintf
('\n************************************************************\n');
2414 %   set(handles.text_win_3_warn, 'Visible', 'on');
2415 %   set(handles.text_win_3, 'Visible', 'off');
2416 %   set(handles.text_win_3_sys, 'Visible', 'off');
2417 %   set(handles.text_win_3_ci1, 'Visible', 'off');
2418 %   set(handles.text_win_3_ci2, 'Visible', 'off');
2419 % else
2420 %   set(handles.text_win_3, 'Visible', 'on');
2421 %   set(handles.text_win_3_sys, 'Visible', 'on');
2422 %   set(handles.text_win_3_ci1, 'Visible', 'on');
2423 %   set(handles.text_win_3_ci2, 'Visible', 'on');
2424 % end
2425
2426 % Display text result
2427 % set(handles.text_win_4_sys, 'String', {round(100 * m(1)) / 100});
2428 % set(handles.text_win_4_ci1, 'String', {round(100 * er_low(1)) / 100});
2429 % set(handles.text_win_4_ci2, 'String', {round(100 * er_high(1)) / 100});
2430 % set(handles.text_win_3_sys, 'String', {round(100 * m(2)) / 100});
2431 % set(handles.text_win_3_ci1, 'String', {round(100 * er_low(2)) / 100});
2432 % set(handles.text_win_3_ci2, 'String', {round(100 * er_high(2)) / 100});
```

```
% Plot the result
set(handles.fig_cover, 'String', []);
set(handles.fig_cover, 'Visible', 'off');
set(handles.fig_result, 'Visible', 'on');
errorbar([1 2], m, (m - er_low), (er_high - m));
set(gca,'xtick',[1 2]);
set(gca,'XTickLabel',{'4R';'3L'});
v = axis;
vdiff = v(4) - v(3);
axis([0.7 2.3, v(3) - 0.07 * vdiff, v(4) + 0.07 * vdiff]);
text(0.83, m(1), {round(100 * m(1)) / 100});
text(0.83, er_low(1), {round(100 * er_low(1)) / 100});
text(0.83, er_high(1), {round(100 * er_high(1)) / 100});
text(2.03, m(2), {round(100 * m(2)) / 100});
text(2.03, er_low(2), {round(100 * er_low(2)) / 100});
text(2.03, er_high(2), {round(100 * er_high(2)) / 100});
ylabel('min.');

% To show the results on a separte window.
% figure;
% errorbar([1 2], m, (m - er_low), (er_high - m));
% set(gca,'xtick',[1 2]);
% set(gca,'XTickLabel',{'4R';'3L'});
% v = axis;
% vdiff = v(4) - v(3);
% axis([0.7 2.3, v(3) - 0.07 * vdiff, v(4) + 0.07 * vdiff]);
% text(0.83, m(1), {round(100 * m(1)) / 100});
% text(0.83, er_low(1), {round(100 * er_low(1)) / 100});
% text(0.83, er_high(1), {round(100 * er_high(1)) / 100});
% text(2.03, m(2), {round(100 * m(2)) / 100});
% text(2.03, er_low(2), {round(100 * er_low(2)) / 100});
% text(2.03, er_high(2), {round(100 * er_high(2)) / 100});
% ylabel('min.');

% End of queuing simulation
```

www.ingramcontent.com/pod-product-compliance
Lightning Source LLC
Chambersburg PA
CBHW081830170526
45167CB00007B/2773